DNAI

The AI Management (AIM) Framework

Best Practices to Integrate AI into your Organizational DNA and Ensure Sustained, Strategic Success

Kartik Sakthivel PhD

Published by:
Brain Fuel Books LLC
PO Box 413
Dover, NH 03820

Interior Design: Kartik Sakthivel
Cover Design: Leslie Sano
Paperback ISBN: 978-1-7377167-7-8
Hardcover ISBN: 978-1-7377167-9-2
Digital ISBN: 978-1-7377167-6-1

ACKNOWLEDGEMENTS

To my parents, P.R. Sakthivel, and Dr. Veena Sakthivel, without whom none of the words in this book could have been possible.

With many thanks to my family, Kiko, Sophie, and Simon, who put up with me being unavailable nights and weekends because I was sequestered in my office working on this book.

Sincere thanks to my friend, coworker, US English grammatical appropriateness evangelist, and fellow "data nerd," Andrew Puente. With a keen eye for consistency and continuity, his exceptional attention to detail has yielded a manuscript that has been meticulously proof-read. Any errors in grammar or punctuation are wholly the responsibility of the author. Thank you, Andrew – hereto forever to be known as "**AI-ndrew**."

To the millions of workers who are on the cusp of an AI revolution – I am rooting for each of you to capitalize on this era and prosper like no generations have before – for the betterment of all humankind.

Table of Contents

PREFACE **xxiii**

Chapter One: The Age of AI **2**

The Need for AI Best Practices *4*

Chapter Two – Demystifying AI **11**

What is AI? *12*

Chapter Three: The Genesis of AI **20**

A Brief History *21*

Chapter Four – Stages, Types, and Branches of AI **30**

Stages of AI *31*

Types of AI *34*

1. Type I – Reactive AI: 34
2. Type II – Limited Memory: 36
3. Type III – Theory of Mind: 38
4. Type IV – Self-Aware: 39

Branches of AI *43*

Chapter Five – An AI Primer: Part 1 **47**

Machine Learning *47*

How Machine Learning Works 48

Machine Learning Methods 51

Chapter Six – An AI Primer: Part 2 **55**

Understanding Neural Networks 55

Neural Networks in the Early 2020s 58

Understanding Deep Learning 60

Chapter Seven – An AI Primer: Part 3 **63**

Understanding Speech AI (TTS and STT) 63

Understanding Natural Language Processing (NLP) 66

Chapter Eight – An AI Primer: Part 4 **70**

Understanding Computer Vision 70

Understanding Robotics 73

Chapter Nine – An AI Primer: Part 5 - Generative AI (A) **77**

Understanding GPT and Large Language Models 78

ChatGPT 79

DALL-E 80

Chapter Ten – An AI Primer: Part 6 - Generative AI (B) **82**

The Promise of Generative AI 82

Generative AI Risks and Challenges 87

KEY RECOMMENDATION ON GENERATIVE AI 91

Chapter Eleven – AI Current State Evaluation **95**

AI Program Maturity (2000-2030) 95

Your Industry's and Organization's Current AI Positioning 101

Factors Influencing AI Adoption Across Industries 103

Why Extrapolate Your Organization's Current AI Positioning *109*

Conducting An AI Current State Evaluation *112*

Question Guide A – Understanding Your Industry's AI Maturity 113

Question Guide B – Understanding Your Company's AI Maturity 114

Chapter Twelve: The AIM Framework© 117

The AIM Framework© *119*

The People/Process/Technology (PPT) Framework 121

Enterprise Best Practice Recommendations 124

People/Process/Technology Best Practice Recommendations 126

Chapter Thirteen: Enterprise Best Practices – Basic Principles: Part 1 129

Basic Principle 1: Organizational Vision {✔ KEY RECOMMENDATION} *130*

Aligning AI with Enterprise Vision 133

Organizational Vision - Expected Value 134

Expected Value 3: Effectuating Cultural Change 135

Expected Value 4: Leadership 136

Expected Value 5: Corporate Governance, Stewardship, and Risk Management 137

"Corporate Allignment": Aligning AI with Enterprise Vision – A CEO/C-suite Guide *137*

"Corporate Allignment" – Twenty Questions for the C-suite 138

Chapter Fourteen: Enterprise Best Practices – Basic Principles: Part 2 143

ii. Basic Principle 2: Communication across the Organizational Hierarchy144

Communications across the Organizational Hierarchy - Expected Value 145

Getting Started 150

When to Start 155

Chapter Fifteen: Enterprise Best Practices – Basic Principles: Part 3 157

iii. Basic Principle 3: Communication across the Organizational Value Chain 158

Communications across the Organizational Value Chain - Expected Value 159

Getting Started 165

Chapter Sixteen: Enterprise Best Practices – Basic Principles: Part 4 173

iv. Basic Principle 4: Continual Measurement, Continual Improvement 174

Continual Measurement, Continual Improvement - Expected Value 175

Chapter Seventeen: Ten Enterprise-Level Best Practices – Part 1 186

1. Enterprise Best Practice 1: Develop an AI Vision, Strategy, and Roadmap at the Departmental Level {⚡ KEY RECOMMENDATION} 187

Building a 360° AI Strategy 188

Chapter Eighteen: Ten Enterprise-Level Best Practices – Part 2 198

Enterprise Best Practice 2: Build vs. Buy Decision {⚡ KEY FINDING/RECOMMENDATION} 200

Tools to Facilitate the "Build" versus "Buy" Decision Process 206

Chapter Nineteen: Ten Enterprise-Level Best Practices – Part 3 218

Enterprise Best Practice 3: Quality Assurance and Quality Control Throughout 219

The Importance of Focus on QA and QC 227

Risks of Inadequate QA and QC 232

Common QA and QC Techniques 233

Chapter Twenty: Ten Enterprise-Level Best Practices – Part 4 **236**

Enterprise Best Practice 4: AI Body of Practice {✓ KEY RECOMMENDATION} *237*

AI Body of Practice – Maturity Model 238

Chapter Twenty-One: Ten Enterprise-Level Best Practices – Part 5 **247**

Enterprise Best Practice 5: Monitoring the Regulatory Landscape {✓ KEY FINDING/RECOMMENDATION} *248*

The Importance of Regulation 248

Regulated Industries 251

Chapter Twenty-Two: Ten Enterprise-Level Best Practices – Part 6 **256**

Illustrative Example – Regulatory Frameworks within the Life Insurance Industry *256*

Regulatory Frameworks to Watch *264*

a. The EU AI Act 265

b. President Biden's Executive Order on AI 272

Chapter Twenty-Three: Ten Enterprise-Level Best Practices – Part 7 **278**

Enterprise Best Practice 6: Careful Vetting of External Data *278*

Enterprise Best Practice 7: Focus on Data Quality and Data Literacy *280*

Enterprise Best Practice 8: Working with Technology Providers *281*

Enterprise Best Practice 9: Clearly Defined Roles and Responsibilities across Enterprise *282*

Enterprise Best Practice 10: Structural Setup for Sustained Success *284*

Chapter Twenty-Four: People/Process/Technology Best Practice Recommendations – Part 1 **288**

People Dimension Best Practices *289*

People Best Practice 1 - Industry Domain Knowledge across Value Chain 290

People Best Practice 2 - Regulatory Knowledge across Value Chain 292

People Best Practice 3 - Dedicated Teams and Resources 294

People Best Practice 4 - AI and ML Governance Model {⚡ KEY FINDING/RECOMMENDATION} 296

People Best Practice 5 - Knowledge Sharing across Enterprise 301

A People-Centered AI Strategy 303

Chapter Twenty-Five: People/Process/Technology Best Practice Recommendations – Part 2 **304**

Process Dimension Best Practices *304*

Process Best Practice 1 - Processes that Promote Transparency and Explainability 305

Process Best Practice 2 - Customer Education on Use of Personal Data 306

Process Best Practice 3 - Documented, Repeatable Data Selection Processes 307

Process Best Practice 4 - Documented, Repeatable Technology Provider Selection Processes 309

Process Best Practice 5 - Testing Rigor, Proactive Publication of Audit Results 310

Technology Dimension Best Practices *310*

Technology Best Practice 1 - Standard IT Supported Technology Stack 311

Technology Best Practice 2 - 3rd Party Data Provider Taxonomy 312

Technology Best Practice 3 - Automated Testing for Inadvertent Bias and Proxy Discrimination 313

Technology Best Practice 4 - Automated QC Processes 314

Technology Best Practice 5 - APIs and Cloud as Core Requirements 315

Chapter Twenty-Six: Fairness and Transparency in AI **318**

Understanding the Need for AI Transparency and Fairness *322*

Mitigating Bias *324*

Chapter Twenty-Seven: Explainable AI (XAI) **329**

The Concept of XAI *329*

Fundamentals and Value of XAI across Industries *332*

Human-Centered and Trustworthy AI *335*

Challenges in XAI Implementation *340*

Effectuating XAI *343*

Chapter Twenty-Eight: Data, Data, Data – Part 1 **349**

Data - So. Much. Data *351*

Explaining Data Governance in the Age of AI *352*

Illustrative Example of a Data Governance Framework: 354

Real-World Data Challenges *361*

Information Sharing 364

Chapter Twenty-Nine: Data, Data, Data – Part 2 **369**

Data in Legacy Systems *369*

Below the Waterline of the AI Iceberg: Equipping Organizations for AI Success *373*

Data Literacy *379*

Definitions 379

Data Literacy Challenges 381

Five Steps to Data Literacy 385

Chapter Thirty – Leading in the Age of AI **389**

Leading an AI-Ready Organization *391*

Chapter Thirty-One – "Hum-AI-n" – Preparing Yourself for the AI Age – Part 1 **404**

AI-Driven Career Changes *404*

Leveraging AI as a Tool *409*

Chapter Thirty-Two – "Hum-AI-n" – Preparing Yourself for the AI Age – Part 2 **414**

Harnessing AI *415*

Chapter Thirty-Three – "Hum-AI-n" – Preparing Yourself for the AI Age – Part 3 **422**

The Softest Skills are the Hardest Skills - Developing Your Soft Skills *425*

Soft Skills in the Age of AI *441*

Chapter Thirty-Four – Applying AI Best Practices – Part 1 **444**

Applying the AIM Framework© - Illustrative Examples *447*

1. Healthcare 449

2. Manufacturing 453

3. Farming and Agriculture: 457

4. Education and Academia: 461

5. The Information Technology Industry: 463

6. Financial Services - Insurance: 467

Chapter Thirty-Five – Applying AI Best Practices – Part 2 **471**

Culture Will Take Special Significance *471*

Applications and Research Extensions of the AIM Framework©: *475*

EPILOGUE A – The End of the Start **484**

The Age of AI *489*

When the Future is Uncertain, Anything is Possible *492*

EPILOGUE B – Sentient AI **496**

The Sakthivel Six-Question AI Sentience Scale © *497*

APPENDIX A: Potential Legal Issues with Generative AI Use **499**

Potential Legal Issues with ChatGPT Use 499

APPENDIX B: LIMRA MarketFacts Article Reprint Permission **506**

Bibliography **507**

Tables of Figures

Figure 1 - AI Event Horizon 4

Figure 2: A Brief History of AI..24

Figure 3: Types of AI and Stages of AI Evolution..30

Figure 4: Stages of AI (encircled) 32

Figure 5: Types of AI 34

Figure 6 – Timeline expectations by when human-level AI will exist. 41

Figure 7: Branches of AI 43

Figure 8: Branches of AI (Simplified)..44

Figure 9: Neural Network (Credit: Free Software Foundation, published under Creative Commons Attribution-Share Alike 3.0 Unported license) 56

Figure 10: AI Program Maturity (in the 21st Century, 2000-2030) 96

Figure 11: Guide A – Industry-Level AI Maturity Preliminary Assessment 114

Figure 12: Guide B - Company-Level AI Maturity Preliminary Assessment 115

Figure 13: The AIM Framework© 120

Figure 14: The AIM Framework© and Supporting Components 121

Figure 15: The People/Process/Technology (PPT) Framework............................124

Figure 16: Enterprise Best Practices Overview 126

Figure 17: The Fifteen People/Process/Technology Aligned Best Practices........127

Figure 18: The AIM Framework© - The People/Process/Technology Best Practices 128
Figure 19: Enterprise Best Practices – Four Basic Principles 130
Figure 20: Enterprise Best Practices: Basic Principles 2, 3, and 4 142
Figure 21: Basic Principle 2: Communication across the Organizational Hierarchy 143
Figure 22: Basic Principle 3: Communication across the Value Chain 157
Figure 23: Recommended Profile for a departmental AI representative 169
Figure 24: Recommended additional support for selected departmental AI representatives. 170
Figure 25: Basic Principle 4: Continual Measurement, Continual Improvement.173
Figure 26: Ten Enterprise-Level Best Practices 186
Figure 27: How the enterprise AI strategy might manifest across the value chain 190
Figure 28: Departmental AI Strategy 194
Figure 29: "Build vs. Buy" decision - some considerations. 199
Figure 30: Technology Vendor Verisk and their Automated/Accelerated Underwriting Workflow (Citation in Body). 205
Figure 31: The Qualitative Build vs. Buy Decision Framework.............................208
Figure 32: The Quantitative Build versus Buy Decision Scorecard.......................213
Figure 33: The Quantitative Build versus Buy Decision Scorecard Scale 217
Figure 34: Simple Supervised Machine Learning Example...................................220

Figure 35: Simple Unsupervised Machine Learning Example 220
Figure 36: QA and QC in Supervised Machine Learning....................................221
Figure 37: QA and QC in Unsupervised Machine Learning 222
Figure 38: Risks and Challenges of Inadequate QA and QC in "Build" and "Buy" Approaches 233
Figure 39: Common QA and QC Methods for "Build" and "Buy" Approaches 234
Figure 40: AI Body of Practice Maturity Model...238
Figure 41: Potential Maturity Dimensions and Quantitative and Qualitative Measures 244
Figure 42: Five Steps to Benchmarking to an AI Maturity Level 245
Figure 43: Examples of Industry-specific Considerations and Importance of Regulatory Frameworks 255
Figure 44: Risk Categories according to the EU AI Act (image credit EY Switzerland) 267
Figure 45: Key Takeaways from President Biden's Executive Order on AI (October 30th, 2023) – credit EY 274
Figure 46: Overview of People/Process/Technology Best Practices...................288
Figure 47: The Vital Need to Focus on People 289
Figure 48: Primary Organizational Components of the AI Governance Triad......300
Figure 49: AI Governance Organizational Framework Overview........................301
Figure 50: The Importance of Well-Defined Enterprise Processes 305
Figure 51: The Synonymous Symbiosis of Technology and AI 311

Figure 52: Causes of Bias in AI systems..319

Figure 53: Methodologies to detect, mitigate, and remediate bias in AI 325

Figure 54: Data Governance Framework used by Uber (Illustrative Example) – credit in text..356

Figure 55: The DAMA Data Governance Framework 360

Figure 56 – Self-reflection questions when considering Data Literacy 381

Figure 57: Ownership of data within an enterprise 382

Figure 58: Impact of poor Data Literacy by the numbers 383

Figure 59: The 1:10:100 Rule of Bad Data 384

Figure 60: Ten Issues Resulting from Poor Data Literacy....................................385

Figure 61: Data Literacy at the base of the "Information Iceberg"......................386

Figure 62: Considerations when building an AI Strategy 392

Figure 63: Planning for Reskilling..399

Figure 64: Driving AI-Driven Reskilling..400

Figure 65: Sakthivel Six-Question AI Sentience Scale 498

PREFACE

"Once a new technology rolls over you, if you're not part of the steamroller, you're part of the road." - Stewart Brand, American author and entrepreneur.

Thanks to the explosive and unprecedented growth of Artificial Intelligence (AI), we are in a world where machines can increasingly think, learn, and make decisions autonomously. Despite its increasing ubiquity across our personal and professional lives, we are still only at the nascent stage of the advent of Artificial Intelligence (AI). AI will continue its accelerated and exponential growth in terms of its complexity, sophistication, autonomy, and maturity; and is poised to fundamentally transform how we work. Through this book, we embark on an exciting journey through the vast landscape of AI. We will explore AI's transformative capabilities, and uncover best practices that enable organizations - regardless of industry and sector - to harness its power effectively, and to ensure we are not steamrolled by this technology.

We will set the stage for learning best practices by first exploring the fundamental concepts of AI. We will examine what AI truly means, what it does NOT mean, its historical development, and its underlying principles. We will also discuss the societal implications of AI. As AI technologies advance, we must carefully consider the ethical, legal, and social dimensions surrounding their deployment.

Understanding the potential risks and pitfalls allows us to navigate this rapidly evolving landscape with caution and responsibility. This is **not** a technical book. It is meant to add clarity and value to business and technology professionals alike. By demystifying the technical jargon and providing clear explanations, I aim to make this book accessible to readers from diverse backgrounds.

As we enter the Age of AI, it is absolutely vital to start off on the right foot with implementation of AI across organizations. There are foundational and fundamental things that companies have to do to find sustained success with their AI strategies and implementations. This book introduces the **Artificial Intelligence Management (AIM) Framework**©. The **AIM Framework**© consists of twenty-five best practices and four basic principles that are outlined in this book. Some of these best practices also present turnkey, customizable and extensible tools that can be applied within your organizations and teams. Apply these AI best practices within your organizations, and I am confident that you will find sustained and long-term success – capitalizing on the promise of AI, while avoiding some of the potential pitfalls. Actioned in unison with your own corporate priorities, the AIM Framework© is a vital prescriptive instrument that guides you to engineer splicing AI into the strands of your corporate DNA in order to equip yourself and your firms to realize sustained and long-term success with AI.

Kartik Sakthivel

I hope you and your enterprises derive value from **DNAI**©.

SECTION ONE

THE AGE OF AI

Chapter One: The Age of AI

"The future is already here – it's just not evenly distributed." – William Gibson (Gibson, 2003).

We are entering an unprecedented era. The 1920s, with less than half of American households having access to electricity, were the age of electrification expanding across America and the world. A hundred years later, the 2020s will be remembered as the age of Artificial Intelligence (AI) expanding across America, and the world.

Each of the first three decades of the 21st century has played host to seminal moments in our history. From a technology perspective, the explosive growth of AI is as seminal a moment as the advent of the internet, and subsequently, the smartphone a few years later. As personal computing picked up in popularity through the 1980s and entrenched itself in the late-1990s, the 2000s will be remembered for the rise of ubiquitous and high-speed internet, the 2010s shall be remembered for the unfettered growth of smartphones and the world's reliance on them, and the 2020s will be remembered for the decade that Artificial Intelligence (AI) becomes our inexorable reality.

Transcending science fiction, whether we realize it or not, AI has become an integral part of our everyday lives. AI's influence is far-reaching, permeating sectors such as healthcare, finance, manufacturing, and entertainment, and leaving no industry untouched. Every industry in every sector is well-positioned to capitalize on the benefits of this expansion of AI – operational efficiencies, automation by immediately disrupting manual and repeatable tasks, business agility, reduced costs, improved customer experiences, and many others.

AI will more than likely lead the charge in terms of all the technology-powered changes across sectors and industries. These changes will bring about the creation and emergence of entirely new industries and professions, fundamentally transform others, and disrupt or disintermediate some. Industries, organizations, professions, and people in these professions that can capitalize on what AI can do for them, would be the ones who realize most success, and minimize their risks and exposure amidst these massive and accelerating changes. From autonomous vehicles navigating our streets to smart assistants residing within our smartphones, AI's impact is reshaping industries, revolutionizing business operations, and propelling us towards new horizons.

And yet, most companies across most established and mature industries are vastly underprepared for it.

The Need for AI Best Practices

As of the first quarter of the 21st century, as a species, across our personal and professional lives – regardless of industries, sectors, companies, and professions – we can consider ourselves to be at the "AI event horizon" of the "AI black hole" in this Age of AI.

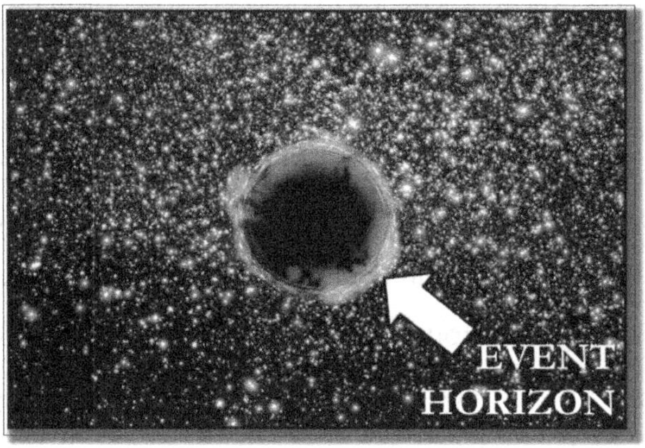

Figure 1 - AI Event Horizon

A black hole and its event horizon are perfect analogies for the Age of AI. A black hole is a region of space where gravity is so strong that nothing can escape its gravitational force, not even light. The event horizon, a sphere surrounding the black hole, is the "point of no return" around a black hole. Anything that encounters the event horizon is unable to escape from being pulled into the black hole. In terms of the Age of AI, we have little insight into what resides inside the "AI black hole", we have no knowledge of what lies on the other side of this "AI black hole", but we can be certain that once we are in the gravitational pull of this black hole and its event horizon, there is no turning back.

The promise of AI is undeniable. It empowers us to solve complex problems, make data-driven decisions, and augment human capabilities. AI algorithms, driven by powerful computational resources, can process and analyze vast amounts of data with unprecedented speed and accuracy. This enables organizations to uncover valuable insights, optimize processes, and gain a competitive edge in an increasingly fast-paced and data-centric world.

Emerging as a transformative force, AI has rapidly permeated various aspects of our lives, heralding a new era of possibilities. In today's rapidly evolving digital landscape, the successful integration and utilization of AI will continue to become increasingly vital for organizations seeking to stay competitive and drive innovation. However, the rise of AI also brings forth a set of challenges and ethical considerations. The decisions made by AI algorithms have far-reaching consequences, impacting individuals, communities, and society at large. If not carefully developed and deployed, AI systems can perpetuate biases, infringe upon privacy, and exacerbate inequalities. Therefore, it is crucial for companies to establish AI best practices that prioritize ethical considerations, transparency, and accountability. Best practices will allow firms to strike the delicate balance between unleashing the potential of AI and ensuring responsible and ethical use.

This book's title, DNAI, is reflective of the premise that in order to harness AI's full potential, incorporating best practices into the very DNA of your organization is of paramount importance. The twenty-five best practices and four basic principles that constitute the **AIM Framework©** encompass a range of turnkey, extensible, and customizable strategies that promote responsible and effective utilization of AI technologies throughout an organization. Implementation of AI best practices as outlined by the **AIM Framework©** offers several significant benefits that shall be reviewed within this book.

We will explore how putting these best practices into action will help organizations avoid some of the challenges that are currently present with AI - and others that will emerge - due to its unprecedented nature. Regardless of the industry you serve in - whether you are an executive in a company, work in a company, run a company, lead a business team, lead a technology team, are a business professional, are a technology enthusiast, or are simply curious about the extraordinary advancements shaping our world - this book of AI best practices aims to provide you with a comprehensive understanding of AI and its practical applications.

One of the key reasons to incorporate AI best practices is to ensure ethical and responsible AI implementation. AI systems are designed to make decisions and perform tasks autonomously, but without proper guidelines and oversight, they can inadvertently perpetuate biases, discriminate, or make incorrect judgments. By leveraging the **AIM Framework©**, organizations can actively address these issues and ensure that AI solutions are explainable, fair, transparent, and accountable. This entails rigorous data governance, regular audits, and ongoing monitoring to detect and rectify any biases or ethical concerns.

Another crucial need for incorporating AI best practices into an organization's AI implementation is to develop trust and reliability in the output of AI systems. When AI technologies are integrated seamlessly into an organization's operations, they can significantly enhance efficiency, accuracy, and decision-making. However, trust is a fundamental element for successful adoption. You have to be able to trust the output from your AI systems, and ergo, the matter of transparency of AI is critical. AI transparency fosters trust in AI. The interpretability and explainability of AI models are a critical need to understand and interpret the decisions made by AI algorithms. Implementing rigorous testing and validation processes, as well as having clear guidelines for handling errors or unexpected outcomes, can instill confidence in stakeholders and users. We will explore strategies to minimize bias, enhance fairness, and ensure AI systems remain aligned with an organization's values and aspirations. By following the **AIM Framework©**, organizations can ensure that AI systems are robust, secure, and reliable.

Great AI + Bad Data = Terrible AI

Data is the lifeblood of AI, and organizations that leverage the **AIM Framework©** can effectively collect, store, and manage data to drive insights and enable informed decision-making. Incorporating the tenets of the **AIM Framework©** will allow organizations to maximize the value of their data. Implementing proper data governance frameworks, including data privacy and protection, and data security measures, ensures that data is handled responsibly and in compliance with relevant regulations. We will explore the bedrock importance of data quality, data literacy, transparency, and fairness, highlighting how these factors influence the explainability, reliability, and trustworthiness of AI systems. By treating data as a strategic asset and establishing mechanisms to harness its potential, organizations can uncover valuable patterns, improve customer experiences, and drive innovation.

As is the case with most large-scale disruptive technological innovations and changes, successfully splicing AI into the very DNA of an organization is a cultural shift. As Peter Drucker has stated, "strategy eats culture for breakfast." A successful exercise to integrate AI as a strategic differentiator into a firm will require the purposeful, intentional, and visionary leadership of an organization's leaders. It will take steady leadership to influence every individual across an organization to actively support and champion this cultural shift. The **AIM Framework©** prescribes best practices that equip leaders to be able to do that.

Returning to where we started, with William Gibson's quote, "The future is already here – it's just not evenly distributed," absent some foundational bedrock principles and best practices, AI will create winners and losers. Companies within an industry, and potentially entire industries, could find themselves stratified based on the maturity of their AI practices, or the number of investments any particular company can afford to make versus others in their field. The **AIM Framework©** is intended to level out the playing field and provide every company across every industry an opportunity to grab the "AI brass ring."

Incorporating AI best practices into the DNA of an organization is crucial for ethical, reliable, and effective utilization of AI technologies. By doing so, organizations can ensure the responsible deployment of AI, foster trust among stakeholders, and maximize the value of their data. The best practices outlined by the **AIM Framework©** are intentionally evergreen in nature. This means that these best practices will stand the test of time, regardless of how sophisticated AI continues to evolve in the future. These best practices will be just as applicable to the AI implementations in 2023 as they will be for the AI implementations in 2053. In the present, as we head into the AI revolution, capitalizing on the **AIM Framework©** will position organizations to be at the forefront of the AI revolution, enabling them to seize opportunities, navigate challenges, and drive sustainable growth in the Age of AI.

Chapter Two – Demystifying AI

The term Artificial Intelligence (AI), although an inexorable part of our lexicon today, is not new – the term dates back to the 1950s. Ask a layperson on their thoughts on AI, and most would cite "*Teradyne Systems*," the "*T-1000*," or "*Skynet*" in response. Most conflate AI with robots and robots with AI. While AI-powered robots certainly do exist and are getting increasingly sophisticated, the personable, household robots as depicted in Isaac Asimov's "*I, Robot*," or the tyrannical and terrifying machines as depicted in the "*Terminator*" series, are - as of 2023 - still a way away. The concept of *Skynet,* however, might actually transcend from science fiction and become science fact much sooner than that. These concerns about the ungoverned growth in the intelligence of AI and the need to establish AI guardrails on a societal level has led to calls for AI regulation by industry experts, something that will be touched upon later in this book. But what exactly ***is*** AI? What exactly ***isn't*** AI? How does it work? And what potential does it hold for the future?

Being able to successfully splice AI into the DNA of your organization – "**DNAI**" - is an exercise in order to engineer the genesis of an organization that thrives in the Age of AI. An enterprise that thrives in the Age of AI is one where AI is simply part of what your organization does, and how it does what it does. Building this "AI-ready," if not an "AI-native" organization, requires a combination of some basic knowledge of AI, as well as a cursory understanding of the most common aspects of AI. We can better position ourselves towards the DNAI journey by sequencing the learning of the basics of AI (in an intentionally non-technical) manner, before exploring the **AIM Framework©**. This chapter, in addition to the two that follow, are intended to ground ourselves into the foundational aspects of AI. This step is necessary in order to gain a fundamental understanding of AI as of the early 2020s, before delving into best practices outlined by the **AIM Framework©**.

What is AI?

In the most basic of definitions, Artificial Intelligence - AI - refers to computer systems capable of performing tasks that typically require human intelligence. These tasks can range from recognizing images and speech, to making decisions, and solving complex problems. AI systems are often connected to each other and/or the internet, and are designed to analyze vast amounts of data, learn from it, and make informed decisions or predictions based on patterns and algorithms.

AI has permeated numerous aspects of our daily lives, often without us realizing it. Virtual assistants like Siri on the iPhone, Amazon's Alexa, and Google Assistant utilize AI to understand voice commands, answer questions, and perform tasks such as setting reminders, playing music, or providing weather updates.

Online streaming entertainment platforms such as Netflix, Max, Amazon Prime Video, Spotify, and shopping sites such as Amazon, Target, Walmart, and Wayfair employ AI to analyze user preferences and behavior, recommending personalized movies, products, music, etc. based on individual tastes. The automotive industry is investing heavily in AI to develop self-driving cars that can navigate roads, detect obstacles, and make real-time decisions to ensure passenger safety. Tesla's Autopilot feature is one of the most prominent examples of this technology.

Consider for a moment that the materials that comprise everything that you see around you have always existed on this planet. It took hundreds to thousands of years for human beings to be able to decipher how to combine these elements together to forge and formulate almost everything manmade that we use today. From the plastics that form your computer keyboard, to the microchips inside your computers, from the electrons flowing back and forth on the fiber optics cables, to airplanes and eyeglasses – everything that constitutes things we take for granted today has always existed. It just takes human ingenuity a long time to find the right combination of these raw elements to create new and innovative materials. In November of 2023, Google DeepMind's new AI tool helped create more than 700 new materials that can be used to make better solar cells, batteries, computer chips, and more. By doing so, DeepMind's AI has found more new materials in a year than scientists have in centuries. Called graphical networks for material exploration (GNoME), the researchers at DeepMind "have trained a deep learning model to predict the structure of over 2.2 million crystalline materials - 45 times more than the number discovered in the entire history of science. Of the two million-plus new materials, some 381,000 are thought to be stable, meaning they wouldn't decompose - an essential characteristic for engineering purposes. These new materials have the potential to supercharge the development of key future technologies such as semiconductors, supercomputers, and batteries, said the British-American company" (Geschwindt, 2023).

"Alongside GNoME, Lawrence Berkeley National Laboratory also announced a new autonomous lab. The lab takes data from the materials database that includes some of GNoME's discoveries and uses machine learning and robotic arms to engineer new materials without the help of humans. Google DeepMind says that together, these advancements show the potential of using AI to scale up the discovery and development of new materials. "While materials play a very critical role in almost any technology, we as humanity know only a few tens of thousands of stable materials," said Dogus Cubuk, materials discovery lead at Google DeepMind, at a press briefing. To discover new materials, scientists combine elements across the periodic table. But because there are so many combinations, it's inefficient to do this process blindly. Instead, researchers build upon existing structures, making small tweaks in the hope of discovering new combinations that hold potential. However, this painstaking process is still very time consuming. Also, because it builds on existing structures, it limits the potential for unexpected discoveries" (Kim, 2023).

AI is revolutionizing healthcare by assisting in early disease detection, medical imaging analysis, drug discovery, and personalized treatment recommendations, all ultimately intended to improve patient outcomes. Using AI, healthcare professionals can analyze vast amounts of medical data, including patient records, imaging data, and genetic information, to improve diagnosis, treatment, and patient outcomes. This includes applications such as predictive modeling for disease progression, identifying high-risk patients, and automating routine tasks like patient monitoring, allowing doctors and nurses to focus on more complex tasks. In other words, the application of AI within healthcare can literally be a life-or-death issue. AI is set to revolutionize the discovery and development of antibiotics and curative treatments. In December 2023, MIT researchers identified a new class of antibiotic candidates using AI. These compounds can kill methicillin-resistant Staphylococcus aureus (MRSA), a bacterium that causes deadly infections. "For the first time in over 60 years, a new class of antibiotics to treat drug-resistant staph infections has been discovered using artificial intelligence (AI) machine learning; a landmark breakthrough to address the antimicrobial resistance

(AMR) crisis. Antimicrobial resistance is a leading cause of death globally, and a public health threat. A projected 10 million will die annually by 2050 due to AMR according to The Review on Antimicrobial Resistance report commissioned by the UK Government" (Rosso, 2023).

A pivotal new era in software development commenced around 2021 with the advent of AI-authored code. Although rudimentary at the time, the advent of OpenAI Codex (OpenAI, 2021) proved transformational on how software was written. In 2016, OpenAI Codex was launched as a platform that translated natural language (such as English statements) into software code. OpenAI Codex - built by OpenAI - a company cofounded by Elon Musk (amongst others), sought to be the world's most ambitious artificial intelligence research laboratory consisting of the for-profit corporation OpenAI LP, and its parent company, the non-profit OpenAI Inc. Lionized for its mission, OpenAI's goal is to be the first to create Artificial General Intelligence, or AGI - a machine with the learning and reasoning powers of a human mind. The lab's goal is to ensure that AI technology was developed safely, and its benefits distributed evenly to the world. The OpenAI charter declares that OpenAI's "primary fiduciary duty is to humanity" (Klok, 2020).

AI-authored software products will imminently necessitate a new way of programming for those entering the computer science field.

As opposed to having to author code for a complex program from scratch, engineers can focus on either tweaking or fine-tuning AI models, or helping AI learn new things. A burgeoning field of study within computer science, in addition to cybersecurity and data science, will emerge within the ethical AI field. The ongoing mission of ethical AI is to ensure AI algorithms continue to benefit humanity and are free from these early days of "Narrow AI", where "clumsy AI" is known to be inadvertently biased and fragile. The "raison d'etre" of companies such as OpenAI is driven by the concern that without the careful guidance of a "benevolent shepherd", which they aspire to be (Klok, 2020), AGI could be catastrophic.

Banks and financial institutions are using AI and Machine Learning (ML) to analyze large datasets, identifying patterns and trends that can help them make more informed decisions around investment and risk management. This includes applications such as fraud detection, credit scoring, and asset allocation. The manufacturing sector is using AI to improve efficiency and quality control, by analyzing data from sensors and other sources to optimize processes and reduce waste. This includes applications such as predictive maintenance, supply chain optimization, and quality control. Companies across industries are using AI to analyze customer behavior, preferences, and feedback to improve customer experience and satisfaction. This includes applications such as chatbots, personalized recommendations, and sentiment analysis.

Unfortunately, most of these industries lack the basic AI best practices they need to ensure they are equipping their firms for long-term AI success. The benefits of AI are vast and far-reaching, with the potential to transform many different industries and use cases. By leveraging these technologies, organizations can improve efficiency, reduce costs, and provide better products and services to their customers. However, absent best practices, these AI implementations would be unable to scale or maintain sustained growth and success. This is where the **AIM Framework©** can serve as an invaluable instrument for companies regardless of industry.

AI will more than likely be leading the charge in terms of all the technology-powered change across sectors and industries. Implicitly, of all the industries and professions that will emerge (and be disrupted) in the next few decades, AI-related industries and professions would likely see strong growth. With less than 40% of American households having access to electricity in most of the 1920s, the 1920s were the age of electrification expanding across America and the world. The 2020s will be remembered as the Age of AI expanding across America and the world. Every industry in every sector is well-positioned to capitalize on the benefits of this expansion of AI – operational efficiencies, automation, business agility, reduced costs, etc. AI will also immediately disrupt manual and repeatable tasks as discussed in the preceding chapters. We haven't even scratched the surface of the tip of the tip of the iceberg when it comes to the utility of ChatGPT and Generative AI – a phenomenon that expanded like wildfire in early 2023.

Chapter Three: The Genesis of AI

While it wasn't until the 21st century, and specifically in the 3rd decade of the 21st century that the accelerating sophistication of AI hit a tipping point, AI, as a concept, is not new. AI has been in varied stages of development since the term was first coined in 1956 at the Dartmouth Conference. For much of the 21st century, we have been using AI extensively across various facets of our lives, without attributing the benefits of this usage to a form of AI. From being able to unlock our smartphones via facial recognition, being able to invoke personal digital assistants such as Siri on the iPhone, and Amazon's Alexa on Amazon's Echo devices, to the relatively unsophisticated algorithms that are pervasive across social media to promote greater human engagement, we have been interacting with AI on a daily basis. On the professional front, the ability for organizations to draw inferences and make predictions based on past performance (predictive and prescriptive analytics) has also been a form of AI. Predictive and prescriptive analytics are built around the central predicate of leveraging algorithms to process data and generate insights, performing computations that would be extraordinarily resource intensive, or nearly impossible, for humans to perform. The development of the COVID-19 vaccine in record time was greatly assisted by the use of AI, demonstrating the value of AI to solve problems that are common across humanity, regardless of nationality, race, or socioeconomic standing.

There have been numerous milestones along the AI journey over these past several decades, some significantly more impactful than others. When the history of AI's growth and maturity is documented over the next several decades, it is highly likely that the year 2023 might well stand out as an impactful one. Generative AI catapulted AI into the mainstream and into our everyday lexicon, making the leap from our personal lives into our professional ones. This in and of itself is at the heart of Generative AI's rapid rise – any technology that catches on in our personal lives and then transitions into our professional lives is significantly more impactful than the other way around. This is the reason why the iPhone was vastly successful, and the Blackberry went extinct, and not the other way around. Propelled by popularity in our personal lives, Generative AI has been able to unlock untapped potential across the corporate value chain with the guarantee of operational efficiencies, human-error reduction, cost savings, new product innovation, faster time-to-market, customer acquisition and engagement, and so on.

A Brief History

AI has been around since the 1950s, when the term was first coined in 1956. Understanding the historical trajectory of AI is essential for navigating its future course. By acknowledging the milestones, challenges, and societal implications of AI development over the decades, we can equip ourselves with the knowledge to shape a future where AI can serve to advance society, while respecting ethical boundaries.

Alan Turing, a British polymath, introduced a concept known as the Turing Test in 1950. In his 1950 paper, Computing Machinery and Intelligence, Turing explored how humans can construct intelligent systems, and how humans can go about measuring their intelligence. The Turing Test, effectively an exploration of a mathematical possibility of artificial intelligence, is a framework to allow determining if a computer system can demonstrate human-level intelligence. The general premise of the Turing Test is that a computer system should theoretically be able to consume available information, render decisions, and solve complex problems just like humans can. According to this framework, if a computer system can engage in communications with humans, without the humans being able to realize that it is a computer system, then the system is said to demonstrate human-level intelligence. The Turing Test is said to be the fundamental stimulus behind the concept and further exploration of AI from that point onwards.

Pivoting from hypothesis to execution and implementation of the Turing Test was quite challenging in the 1950s. Computers in this decade lacked the key requirement for developing intelligence, which is the ability to store information. Computer systems in the early 1950s could execute commands, but did not have memory.

They were also extremely expensive, with leasing a computer costing at least $200,000 per month. This meant that only large technology companies, well-funded universities, and the Federal Government could afford to experiment and ideate with these systems. Rockwell Anyoha summarizes the limitations of the 1950s compared to the 21st century the best in a 2017 blog post on Harvard University's Graduate School of Arts and Sciences site, stating, "We haven't gotten any smarter about how we are coding artificial intelligence, so what changed? It turns out, the fundamental limit of computer storage that was holding us back 30 years ago was no longer a problem. Moore's Law, which estimates that the memory and speed of computers doubles every year, had finally caught up and in many cases, surpassed our needs. This is precisely how Deep Blue was able to defeat Gary Kasparov in 1997, and how Google's Alpha Go was able to defeat Chinese Go champion, Ke Jie, only a few months ago. It offers a bit of an explanation to the roller coaster of AI research; we saturate the capabilities of AI to the level of our current computational power (computer storage and processing speed), and then wait for Moore's Law to catch up again" (Anyoha, 2017).

In 1956, John McCarthy of Dartmouth College and Marvin Minsky of the Massachusetts Institute of Technology (MIT) hosted the "Dartmouth Summer Research Project on Artificial Intelligence" (DSRPAI) along with Nathaniel Rochester of IBM and Claude

Shannon of Bell Laboratories, at Dartmouth College in Hanover, New Hampshire. This conference was where the term "Artificial Intelligence" was first used, and it led to the founding of AI as a research discipline. It was at this conference where the "Logic Theorist" was first presented. Logic Theorist was a computer program created by Allen Newell, Cliff Shaw, and Herbert Simon, and was funded by the (Research and Development) RAND Corporation. Widely considered to be the first "artificial intelligence" program, Logic Theorist was designed to mimic the problem-solving skills of a human. Figure 2 below depicts a brief history of AI through the decades.

Figure 2: A Brief History of AI

AI experimentation and innovation took off in the 1960s as computing power continued to increase and become more cost effective. By the 1960s, computers could not only process instructions, but also store information. In addition to the ability to store data, relatively economically, computer systems also continued to become faster and a lot more accessible. This effectively democratized investments and innovation in AI, since AI advancements did not need to solely rely on well-funded educational institutions or the Federal Government. In 1965, Joseph Weizenbaum, a German American computer scientist and MIT professor (and after whom The Weizenbaum Award is named), developed an AI system known as ELIZA. ELIZA can be considered an early precursor of the modern chatbot, with the capability to interpret spoken language. ELIZA is said to be ground-breaking in that it built the foundation for modern and sophisticated conversational AI systems, such as Siri and Alexa.

DENDRAL, a rule-based expert system, was introduced in 1972 and is renowned as the first of its kind. DENDRAL (for Dendritic Algorithm) was a computer program devised by Geneticist Joshua Lederberg, Edward A. Feigenbaum, who was chairman of the Stanford computer science department, and Carl Djerassi, a chemistry professor. DENDRAL ran on a computer system called ACME (Advanced Computer for Medical Research), installed at Stanford Medical. Designed to relieve chemists of a task that was challenging, tedious, and time-consuming, the goal of DENDRAL was to help in the explanation of the molecular structure of unknown organic compounds taken from known groups of such compounds. Once fully operational, DENDRAL performed these monotonous and repetitive tasks with greater speeds than and with comparable accuracy to human experts. DENDRAL's greatest contribution was to the development of knowledge-based AI transferring the principles of AI "from the realm of chess and other strictly controlled settings in which they had been formulated during the 1950s, to real-world problems facing biomedical researchers and physicians. They wanted to show that computers could become experts within a concrete knowledge domain, such as mass spectrometry, where they could solve problems, explain their own conclusions, and interact with human users" (NIH, 2019).

Significant strides were made in AI research in the 1960s and 1970s, with the development of expert systems capable of solving complex problems in specialized domains. However, limitations in computational power and the inability to handle uncertainty hindered further progress. The subsequent AI winter in the 1980s saw a decline in enthusiasm and funding due to unmet expectations and overpromising. Nonetheless, this period fostered essential introspection and laid the groundwork for the resurgence of AI. Similar to the limitations AI development faced in the 1950s, computers in the early 1980s, could not store enough data, or process it fast enough, and simply lacked the sheer computational power to do anything that could be considered significant in terms of AI. This turned around towards the mid to late 1980s, and "AI was reignited by two sources: an expansion of the algorithmic toolkit, and a boost of funds. John Hopfield and David Rumelhart popularized "deep learning" techniques which allowed computers to learn using experience. On the other hand, Edward Feigenbaum introduced expert systems which mimicked the decision-making process of a human expert. The program would ask an expert in a field how to respond in a given situation, and once this was learned for virtually every situation, non-experts could receive advice from that program. Expert systems were widely used in industries. The Japanese government heavily funded expert systems and other AI related endeavors as part of their Fifth Generation Computer Project (FGCP). From 1982-1990, they invested $400 million dollars with the goals of revolutionizing computer processing, implementing logic programming, and improving artificial intelligence." (Anyoha, 2017).

AI development continued to accelerate in the 1990s as computer systems became increasingly sophisticated, capable of increased amount of storage and faster processing times. IBM's Deep Blue defeated world chess champion Garry Kasparov in 1997, a symbolic victory of AI over human intelligence. By the late 1990s, the rise of the internet changed the AI landscape, and the ubiquitous nature of the internet into the early 2000s, along with exponentially increasing internet speeds, made access to global data significantly easier. The 21st century has heralded a new era of AI. Powered by machine learning techniques, particularly deep learning, breakthroughs in neural networks, fueled by vast datasets and enhanced computing capabilities, have revolutionized the field. Applications ranging from image and speech recognition to natural language processing have surged, reshaping entire industries and societal norms. Widely renowned as the "Godfather of AI", Geoffrey Hinton, brought deep learning into the forefront in 2006. Deep learning served as the catalyst for increasing AI sophistication and innovation in the early 21st century. Ian J. Goodfellow, American computer scientist, engineer, and executive, incepted Generative Adversarial Networks (GANs), a ML framework for Generative AI around 2014, which was closely followed by the creation of OpenAI in 2015.

It is helpful to understand the historical landscape of AI for several reasons. Firstly, the cyclical nature of AI development thus far, marked by periods of excitement followed by disillusionment, highlights the importance of learning from the past. Acknowledging the challenges faced in the past aids in making informed decisions for future advancements, and breathless excitement about the next AI innovation should be tempered by continually measuring the value of ongoing AI implementations. Recognizing the evolution of AI illuminates the transformative journey from symbolic AI to contemporary AI, and provides a context to appreciate the evolution of technologies, and the underlying theories that have culminated in contemporary AI capabilities. Delving into the history of AI unveils ethical dilemmas and societal impacts that have arisen along the way, and understanding historical contexts can mitigate potential risks and foster responsible AI practices, as well as the ethical development and deployment of AI systems. Having historical perspective can serve as some kind of a “true north” of inspiration for future AI innovators. Insights gleaned from past breakthroughs and challenges can trigger new ideas, driving innovation in AI research and application domains.

Chapter Four – Stages, Types, and Branches of AI

We have a tendency to clump a wide range of technology that provides predictive capabilities, or exhibits some level of self-learning or intelligence, as "AI." Whereas this overarching terminology is generally accurate, it is important to distinguish the taxonomy between the different types of AI, where AI currently is in its evolution, where it is likely headed, and have a cursory understanding of the plethora of "branches" of the "AI tree."

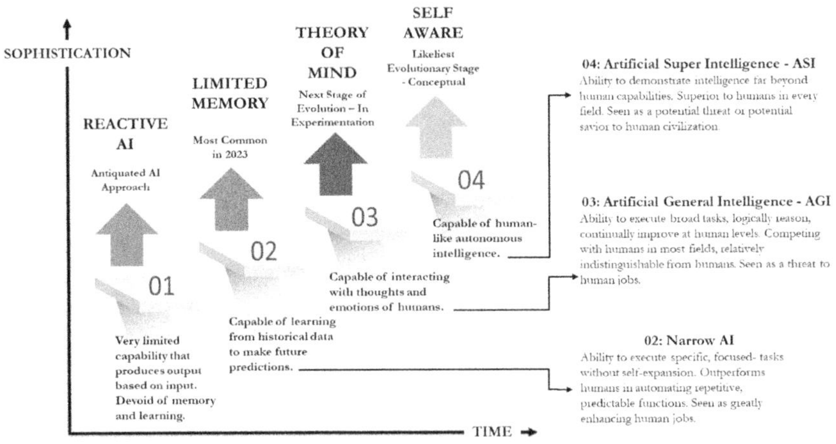

Figure 3: Types of AI and Stages of AI Evolution

Figure 3 visually represents the evolution of the AI field, with each of the three evolutionary stages mapped to the four main types of AI as are defined (as of 2023). The four types of AI are graphed to the increasing passage of time on the X-axis and the increasing sophistication and maturity of AI on the Y-axis. AI is a vibrant and rapidly evolving field. The terminology used in the first half of the 21st century's third decade might be different in the latter half. The categories of AI – the AI Types – are closely linked to how the AI field has evolved thus far and therefore Figure 3 depicts both, the stages of AI Evolution, as well as the corresponding types of AI. Figure 3 will be the predicate for further exploration of the types of AI (categories) and teasing apart each stage of AI evolution. Despite the relative fluidity of the field, the goal of the **AIM Framework©** is to provide an extensible and scalable AI governance toolset that can stand up to the ongoing evolution of AI as a sprawling industry.

Stages of AI

There are two fundamental stages of AI - Narrow AI and General AI, with an emerging third stage of AI evolution that is a derivative of General AI, known as Super Intelligence. Figure 4 presents an overview of the Stages of AI (encircled portion of the figure). These classifications of AI exist as of 2023, and although there is broad consensus around the two main stages of Narrow AI (also known as Weak AI) and General AI (also known as Strong AI), there is an increasing desire to stratify AI stages into a third category known as Super Intelligence in addition to the two primary ones. ASI is a relatively new category of the Stages of AI.

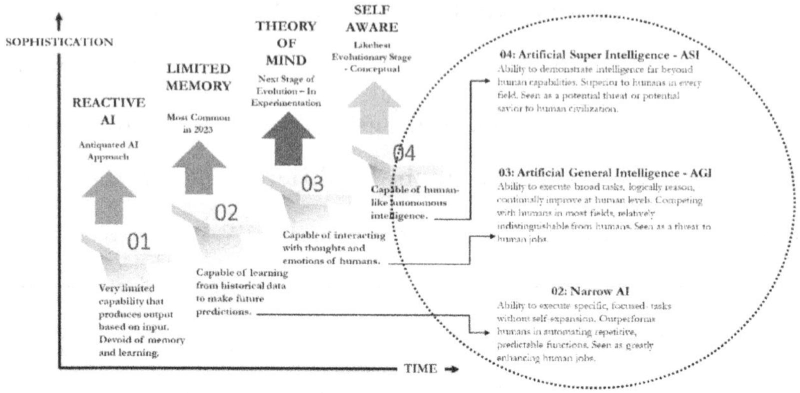

Figure 4: Stages of AI (encircled)

1. **Narrow AI (Artificial Narrow Intelligence – ANI)**, also known as **Weak AI**, is designed to perform very specific tasks with a high level of proficiency. In business parlance, one would say that Narrow AI provides depth, not breadth. These specific tasks include speech recognition, image classification, recommendation systems, and even autonomous vehicular operations. Narrow AI systems excel in their specialized and specific domains, but lack the ability to generalize beyond their predefined tasks.

2. **General AI (Artificial General Intelligence – AGI)**, also known as **Strong AI**, represents the concept of machines - computers - possessing human-like intelligence across a wide range of tasks. AGI systems are those that can exhibit self-

awareness, what we call “consciousness”, and the ability to understand and learn any intellectual task that a human can. Some equate AGI with *sentience*. However, achieving AGI remains a challenge, and current AI systems are many years away from achieving this level of complexity and sophistication.

3. A third category, **Artificial Super Intelligence**, or **ASI**, is a derivative of AGI. ASI is hypothesized to become a reality not too long after AGI is achieved. With ASI, computers are projected to demonstrate intelligence that is far beyond human capabilities. ASI is seen as becoming superior to humans in every sector and in every field. ASI has also been the subject of much debate and discussion in the United States Congress (as of 2023). Calling ASI an existential threat to humanity, elected officials and private industry leaders have been deliberating how a public/private partnership can realize the value of ASI to help humankind versus the inherent risk to humanity that it can pose.

Types of AI

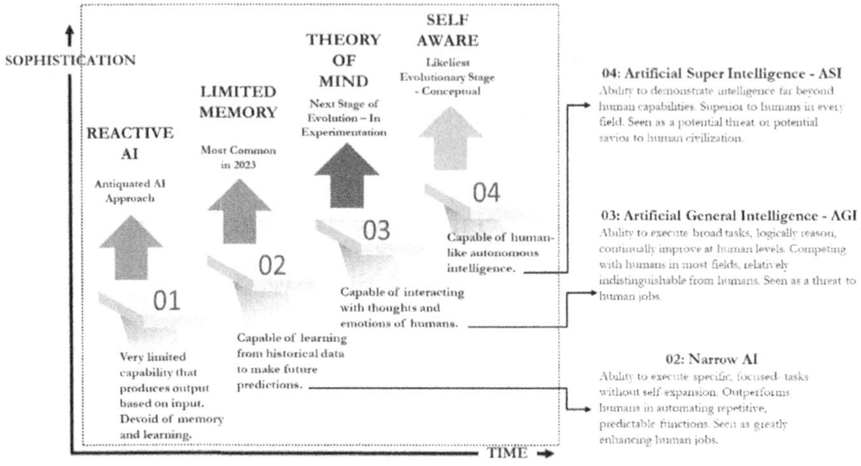

Figure 5: Types of AI

There are four primary types of AI that map to the three evolutionary stages, as stratified in 2023. The four types of AI are based on the sophistication and maturity of AI over an increasing length of time. Figure 5 outlines these four types (represented within the boxed enclosure). Examining these four types a bit further:

1. Type I – Reactive AI:

Reactive AI is the least sophisticated form of Artificial Intelligence. Reactive AI is predicated on deterministic algorithms, so it “reacts” in a relatively predictable manner in response to a pre-programmed set of directives. Operating within the parameters of a set of predefined rules, Reactive AI lacks the ability to know of any of its past experiences, or learn from these cumulative past experiences. Focused on executing predefined responses to specific inputs, Reactive AI cannot adapt and adjust by learning historical data over time.

Any repeatable, rote, operational processes – with a defined, limited, or predictable set of inputs and/or a defined, limited, or predictable set of outputs – can be automated with Reactive AI. An example of Reactive AI in practice includes factory robots along an assembly line. These robots are programmed with specific instructions to executive specific tasks. They lack the capacity to learn from their past in order to make concerted operational improvements for the future. They do not learn from their historical assembly patterns to exercise judgment in bringing forward any operational improvements. They are programmed to carry out specific assembly line functions, which they execute with predictable outcomes. These outcomes are often binary in nature – whether the Reactive AI succeeded in executing its instructions or it failed. Another example would be a vending machine. No matter how sophisticated a vending machine is, the job of the vending machine is to respond to a limited number of inputs – an alphanumeric choice selected by a consumer – in order to dispense a specific product associated with this alphanumeric choice. The vending machine does not know about the consumer's historical purchase patterns, or any other facts about the consumer, in order to assist with making a product selection.

2. **Type II – Limited Memory:**

As the name suggests, with the Limited Memory type of AI, AI is said to possess a limited amount of short-term memory. AI implementations categorized as Limited Memory, typified by their ability to look into the past and improve over time, are the most pervasive implementations of AI as of 2023. In the overall scheme of AI's current and upcoming evolution, Limited Memory AI implementations are relatively quite mature as of 2023. These AI implementations have the ability to store and leverage recent data, but lack the ability to store, recall, and leverage longer-term historical data. While Limited Memory AI implementations have the capability to learn from the past, they can only make decisions based on relatively recent experiences. Limited Memory AI implementations are capable of learning from historical data in order to make future predictions.

Limited Memory AI is currently pervasive across three major ML model types:

i. *Reinforcement Learning* – AI systems that learn, and continually learn, through repeated trial and error.

ii. *Long Short-Term Memory (LSTM)* – AI systems that learn from past experiences in order to make predictions about the future.

LTMS AI systems leverage the totality of historical data available to them, giving more weightage to the most recent data, to make decisions based on the entire data set, and render predictions about what will happen next.

iii. *Evolutionary Generative adversarial networks (E-GAN)* – AI systems where the AI/ML model continually seeks out better paths to make decisions. These systems use statistics and conduct countless simulations in the blink of an eye to predict outcomes. These systems evolve over time as they continue to literally and figuratively blaze new trails to arrive at the optimal predictions in an optimized, resource-efficient fashion.

Stylized to how neurons function in the human brain, systems equipped with Limited Memory AI possess decision-making capabilities using sophisticated classification, pattern recognition, pattern matching, and historical data referencing. Using the same historical data, Limited Memory AI can render inferences using historical data (business intelligence), and leverage the same data to make predictions (predictive intelligence).

Almost all instances of Limited Memory AI use the same six steps to facilitate Machine Learning:

i. The AI/ML model has to be developed. The model can be developed entirely by humans or by humans assisted by other systems. *When computers gain the capabilities to autonomously develop AI/ML models, the evolution of AI would have progressed to the next level. Limited Memory AI is categorized by the fact that machines, as of 2023, do not possess this capability.*

ii. The AI/ML model should have the capacity and capability for making predictions.

iii. Training data must be supplied to this AI/ML model.

iv. The model needs to have the ability to accept feedback data from humans and/or the environment in which it operates.

v. The AI system should have the ability to store this feedback data.

vi. The model should have the ability to incorporate this feedback data into its operations to render decisions and/or make predictions the next time that it runs (learning capability based on feedback that is provided by humans and/or other systems and/or the environment).

Limited Memory AI systems comprise the predominant majority of AI implementations as of 2023, including increasingly sophisticated chatbots, virtual assistants, navigation systems, personalization recommendation engines within streaming media platforms, ecommerce website shopping recommendations, self-driving vehicles, etc. Limited Memory AI has already added significant value in exponentially increasing the reaction time of self-driving vehicles.

3. Type III – Theory of Mind:

Theory of Mind AI is a sophisticated, and as of 2023, a type of AI that is in various states of experimentation maturity. The general principle behind Theory of Mind AI is an AI that can understand, comprehend, decipher, infer, and consequently predict the emotions, intent, and beliefs of humans. Theory of Mind AI aims to "independently" think about the feelings of humans, if not outright give the impression that it "feels" those feelings. Theory of Mind AI is a step towards AI that emulates the innate human qualities of understanding emotions and expressing empathy, thereby enabling creation of systems that interact more naturally and effectively with humans.

There are innumerable uses cases for Theory of Mind AI to add immense value. Imagine having AI powered chatbots that can sense dissatisfied customers and change their approach to deescalate or diffuse interactions with an irate customer. Another valuable implementation to consider would be within virtual assistants such as the Amazon Alexa, iPhone's Siri, or Google Assistant that could analyze, interpret, and comprehend a human's mood through tone, voice modulation, facial expressions, and adjust its nuanced interactions appropriately.

4. Type IV – Self-Aware:

As the name suggests, Self-Aware AI is AI that is said to have *consciousness*. A self-aware AI recognizes its own existence. It knows of "who" it is and "what" it is. The AI system is cognizant of its own capabilities and limitations. Self-Aware AI is still outside the technological capabilities of AI as of 2023. A paper published in 2018, several years before the explosion of Generative, in the Journal of Artificial Intelligence Research states "Advances in artificial intelligence (AI) will transform modern life by reshaping transportation, health, science, finance, and the military. To adapt public policy, we need to better anticipate these advances. Here we report the results from a large survey of machine learning researchers on their beliefs about progress in AI. Researchers predict AI will outperform humans in many activities in the next ten years, such as translating languages (by 2024), writing high-school essays (by 2026), driving a truck (by 2027), working in retail (by 2031), writing a bestselling book (by 2049), and working as a surgeon (by 2053). Researchers believe there is a 50% chance of AI outperforming humans in all tasks in 45 years and of automating all human jobs in 120 years, with Asian respondents expecting these dates much sooner than North Americans" (Grace, Salvatier, Dafoe, Zhang, & Evans, 2018). Just based on how ubiquitous AI is increasingly becoming, it is likely that some of the dates the authors postulated are actually occurring earlier than anticipated. For instance, the authors speculated that AI would be capable of writing high-school essays by 2026. Thanks to OpenAI's ChatGPT, this date shifted into 2023.

According to an article published in February of 2023 by Max Roser, 356 AI experts were asked when they believe there would be a 50% chance that human-level AI (defined as unaided machines being able to accomplish every task better and more cheaply than human workers) would exist. According to this article, which references the study by Grace et al above, and is encapsulated in Figure 6, "half of the experts gave a date before 2061, and 90% gave a date within the next 100 years" (Roser, AI timelines: What do experts in artificial intelligence expect for the future?, 2023)

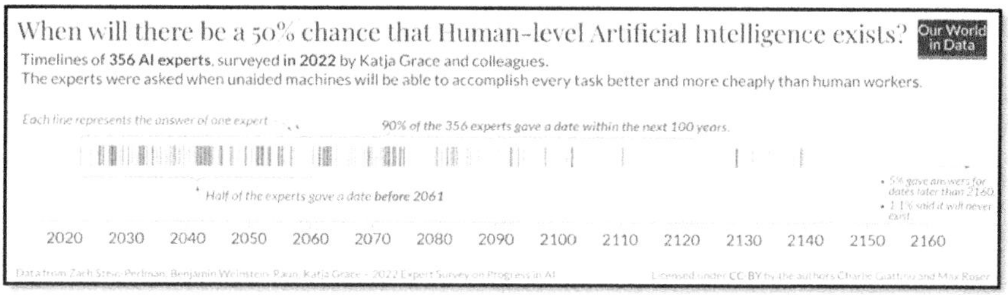

Figure 6 – Timeline expectations by when human-level AI will exist.

While one could make any kinds of predictions over what will transpire within a century since it is unlikely that the predictors themselves would be around to be held accountable even if they are spectacularly wrong, the general idea that Self-Aware AI will exist in some form within a hundred years is perfectly plausible. In fact, if

American inventory and futurist Ray Kurzweil's predictions ring true, 2045 will be the year that humankind would achieve "Technology Singularity." Kurzweil has made almost 200 futuristic predictions throughout his career and had an impressive accuracy rate of close to 90%. He has posited 2045 to be the year that "Singularity" would be achieved, by defining this moment of "Singularity" as - "which is when we will multiply our effective intelligence a billion-fold by merging with the intelligence we have created" (Reddy C. , 2017). In other words, Kurzweil has predicted that by 2045, humans would integrate intelligent technology within our bodies – or in sci-fi terms, merge with intelligent machines – to boost our intelligence and dramatically improve our quality of life. While not quite Self-Aware AI in the true, standalone sense, it is the augmentation of human intelligence with artificial intelligence that might serve as an intermediary step towards fully Self-Aware AI systems.

Self-Aware AI exhibits a level of consciousness and introspection and can make decisions beyond its original programming. Self-Aware AI concepts cover different levels of AI capabilities. These range from basic rule-based reactions, to advanced understanding of others, and ultimately, self-awareness. These categories represent a progression from simple rule-based reactions to more sophisticated AI that can learn, understand others, have the capacity to understand human-like mental states, and develop self-awareness, including a form of artificial consciousness. As of 2023, Self-Aware AI remains theoretical in practice. Once the realm of science fiction, given the rate of AI progress, this might well become science fact much sooner than anticipated. It is also one of the primary drivers behind technology executives raising the alarms with elected officials regarding the very real perils of AI that lacks parameters and governance - these concerns being predicated on their assessments that AI poses an existential threat to human civilization.

Branches of AI

While Generative AI (GAI), and specifically ChatGPT, might be recognized over history as the application of AI that served as a catalyst to usher in the Age of AI, it is important for business and technology professionals to clearly understand that AI, as a field, is **NOT** a monolith. AI is a sprawling field with a multitude of branches and specializations that all generally derive from the same basic fundamentals (all have some kind of AI models, training data, input data, and output).

Figure 7: Branches of AI

There are a myriad of ways experts have tried to sketch the continually and rapidly growing "AI tree". Appearing in "Investigating the Influence of Artificial Intelligence on Business Value in the Digital Era of Strategy: A Literature Review" by Nikolaos-Alexandros Perifanis and Fotis Kitsios, Figure 7 (Perifanis, 2023) presents one such depiction of the branches of the "AI tree."

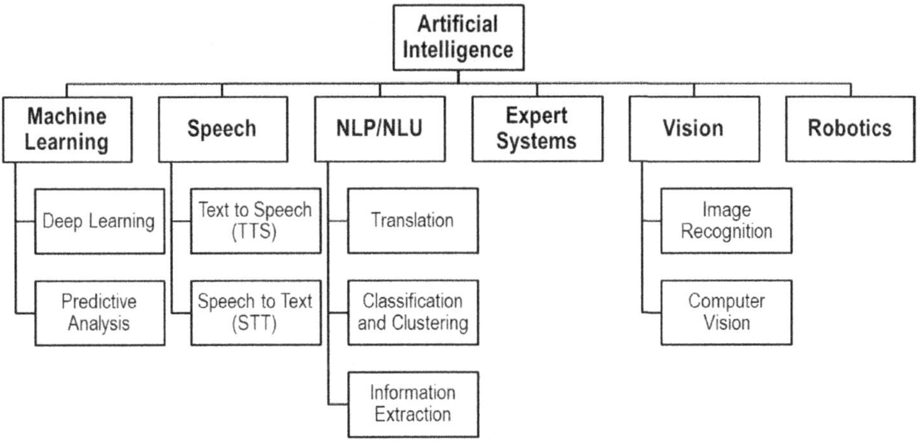

Figure 8: Branches of AI (Simplified)

Depicting the field of AI into branches as represented in Figure 8 will suffice for the purposes of providing the broadest (and simplest) overview of AI possible. Section Two of this book provides a non-technical and easy to understand explanation of most of these branches as part of six "AI Primer" chapters.

"Section Two – An AI Primer" encompasses six chapters that explain several AI fields and their applications in an easy, business-friendly manner.

If you are already adequately AI-savvy, you can skip to "Section Three – AI BEST PRACTICES."

SECTION TWO

AN AI

PRIMER

Chapter Five – An AI Primer: Part 1

The next six chapters will present a fundamental overview of some the basic fields across the sprawling AI ecosystem. The central predicate for most AI implementations starts with the concept of Machine Learning (ML), which - at its core - consists of the "ML model," and the underlying data that flows through the model. The underlying data itself can be broadly categorized into two – the "training data," that is the data that has been used to train the model into achieving a desired outcome, and the actual/real/production data that an AI/ML model uses to generate its output. The output itself could range from a prediction, an image, video content, etc.

Machine Learning

At the heart of AI is a concept known as Machine Learning, or ML for short. The goal for ML is to develop computers/machines that can learn and adapt, just like people do. ML envisions machines having the ability to learn from their experiences, make sense of information, and make decisions based on what they have learned.

ML is a facet of AI which focuses on enabling machines to self-learn from vast amounts of data. At its core, ML is about enabling computers to learn from examples and experiences, rather than being explicitly programmed for every task. ML allows machines to have the power to teach themselves, and improve their performance over time. ML can be considered as a way to give machines the ability to learn and grow on their own, analogous to a human child discovering the world around it.

ML is possible due to "ML algorithms", which are software code and programs. The programming behind these algorithms can be complex, and these algorithms can be combined with others to create extraordinarily complex and sophisticated software programs. The premise of ML is that machines continue to self-learn and render decisions based on their programming, improving their own performance and accuracy over time. ML algorithms are intended to execute autonomously, without the need for any additional and explicit software coding or programming. ML algorithms need to be "trained" in order to arrive at an acceptable level of being able to ingest data, and make predictions or recommendations based on that data. These ML algorithms are therefore trained on vast datasets. These massive datasets allow ML algorithms to recognize patterns and make predictions or decisions based on that acquired, "learned" knowledge.

How Machine Learning Works

As stated earlier, **Great AI + Bad Data = Terrible AI**. AI and ML all start with the data. ML consumes vast amounts of data. This is the information that machines can use to learn patterns and make predictions. This data can come in many forms such as text, images, and numbers. ML can also ingest data in the form of sensory inputs like sound or touch. At the highest level, ML operates in three steps – input, learning, and output.

Input is the first step in Machine Learning. In this step, data is fed into a learning algorithm. As discussed earlier, an algorithm is a set of programmed instructions that serves as a guide to the machine's learning process. It's like a recipe that tells the computer how to make sense of the data and extract meaningful information from it. Next, the machine analyzes the data, searching for patterns and relationships that might not be immediately apparent to human observers. It looks for similarities, differences, and trends within the data, seeking to understand the underlying structure and make sense of the information it contains. As the machine continues to analyze the data, it adjusts its internal model. The internal model is the machine's internal representation of what it has learned through the present time. This model acts as the machine's understanding of the data and forms the basis for its future predictions or decisions. The machine learns by comparing its predictions or decisions with the actual outcomes or correct answers. It receives feedback on its performance, allowing it to refine its understanding and improve its accuracy over time. It is like a continuous feedback loop, wherein the machine learns from its mistakes and adjusts its internal model accordingly.

A simple example is useful to help illustrate and understand ML's underlying process of input-learning-output. Consider you have an AI-powered box of crayons. These AI crayons are able to draw pictures based on what you simply ask them to draw. At first, the AI crayons do not know how to draw anything, but with little knowledge, their ability to draw improves over time. Following the three steps of ML, in ML, everything commences with the input step. This step is akin to providing the AI-powered crayon box with something to learn from. This input could be a collection of pictures, numbers, or even text. For instance, if we want the AI crayons to draw dogs, we would show it many pictures of dogs as input. In the second step, learning, once the AI crayons have had their input, it is time for them to learn by using algorithms (a set of instructions) to analyze the input and find patterns. The AI crayon box looks for similarities between different pictures of dogs, such as the shape of their ears, the way their eyes look, size, relative length of their coat, and other distinguishable attributes. The more pictures of dogs that the AI crayon box sees, the more it starts to notice patterns. It becomes increasingly better at recognizing what makes a dog a dog, and starts to recognize the shape of a dog's snout, the tail, ears, and other pertinent distinguishable attributes.

The AI crayon box learns from its mistakes and adjusts its understanding based on the feedback it receives. Comparable to when humans practice a new skill, the AI crayon box improves over time. In the final phase, output, the AI crayon box has to show what it has learned. When you ask the AI crayon box to draw a dog, it looks at the features it learned from the input, combines them, and draws a picture that resembles a dog. Sometimes, the AI crayon box might make mistakes, but with more practice and feedback, it gets better and better at drawing distinct types of dogs. In the real world, this output could be predictions, decisions, or even creative works like art or music. Machines use what they have learned to make educated guesses or perform tasks based on the patterns they have discovered in the input.

Machine Learning Methods

But how does the AI crayon box learn? It uses a type of ML called "Supervised Learning", or “Supervised Machine Learning”. Supervised Machine Learning is comparable to having a teacher or a coach guiding you through the learning process. Continuing our simple example from above, the teacher is a person who tells the AI crayon box whether its drawings are correct or not. This feedback helps the AI crayon box to adjust its understanding and get better at drawing dogs. There's also another type of Machine Learning called "Unsupervised Learning" or “Unsupervised Machine Learning”. In Unsupervised Machine Learning, a machine can learn on its own without a teacher. It does so by looking for patterns in the input and organizes the information in a meaningful way, akin to exploring a new city without a map and discovering interesting places on your own. Supervised Machine Learning and Unsupervised Machine Learning are the two main types of ML. Both approaches have their strengths and are used in various real-life scenarios. A third type known as “Reinforcement Learning” is also common, although less so than Supervised and Unsupervised.

1. *Supervised Machine Learning*: In Supervised Machine Learning, models are trained using labeled examples. These examples consist of input data (such as images or text) and their corresponding correct outputs. The algorithm learns to associate inputs with outputs, enabling it to make predictions on new, unseen data. In Supervised Machine Learning, the machine receives feedback on whether its predictions are correct or not, allowing it to adjust its understanding and refine its predictions. Supervised Machine Learning is often employed in tasks like image recognition, language translation, and speech recognition.

2. *Unsupervised Machine Learning*: Unsupervised Machine Learning involves training models on unlabeled data. The algorithm explores the data, searching for patterns or structures within it. This approach is particularly useful for tasks like clustering similar data points or finding hidden patterns in large datasets. Unsupervised Machine Learning is like a curious explorer navigating through a world of unknowns. Without any explicit guidance, the machine searches for patterns and structures in the data. It groups similar

things together and identifies interesting relationships without being told what to look for. It's like discovering hidden treasure in an uncharted territory. Unsupervised Learning shines in tasks such as customer segmentation, anomaly detection, and data clustering.

3. *Reinforcement Learning*: Reinforcement learning employs a reward-based system. The algorithm learns by interacting with an environment and receiving feedback in the form of rewards or penalties. It aims to maximize the rewards by taking actions that lead to positive outcomes, thus enabling the machine to learn through trial and error.

Machine Learning is revolutionizing the way we solve problems and make decisions. From voice assistants like Siri and Alexa, to personalized recommendations on streaming platforms, to predicting weather patterns, detecting fraudulent activities, powering self-driving cars, etc., machines are becoming increasingly skilled at understanding and interpreting complex data due to ML. However, ML also has inherent challenges. Availability of quality data continues to be a major challenge. ML algorithms rely on vast amounts of data to learn effectively. Ensuring that the data is diverse, representative, and of high quality is crucial for accurate and unbiased learning. Another challenge is the issue of bias. Machines learn from the data they are provided, and if that data contains biases or prejudices, it can lead to biased predictions or decisions. It's essential to be mindful of this and actively work towards building fair and ethical machine learning systems. Data privacy and security are also critical considerations. As machines learn from vast amounts of personal data, it's important to safeguard that data and ensure that it is used responsibly and in compliance with privacy regulations. Ensuring that machines make fair and ethical decisions, protecting data privacy, and avoiding biases are crucial considerations as ML advances and becomes more sophisticated.

As AI continues to advance, we can expect Machine Learning to become even more powerful and pervasive. ML will continue to evolve rapidly with the development of advanced machine learning algorithms, the integration of ML with other branches of AI like robotics and Natural Language Processing (NLP), and the exploration of new horizons in areas like Reinforcement Learning and Deep Learning. ML is a powerful tool that empowers machines to learn, adapt, and make intelligent decisions.

Chapter Six – An AI Primer: Part 2

To fully get a basic understanding of AI, it is important to gain an appreciation for terms such as "Neural Networks," "Deep Learning," "Natural Language Processing," etc. This chapter will continue a high-level overview of the facets of AI. It is important for business and technology leaders to gain this basic understanding in order to have a demystified view into AI, in advance of delving into best practices.

Understanding Neural Networks

Neural Networks are the building blocks of advanced Machine Learning. Neural Networks can be thought of as a vast network of interconnected neurons that are structurally similar to the neurons that comprise a human brain. Each of these interconnected neurons have special abilities. In a human brain, each neuron receives information, processes this information, and then passes this information along to other neurons within this vast interconnected network of neurons. These neurons work together, forming a complex web of connections that enables the human brain to process information and decipher the world around it. In the same way, Neural Networks in the world of AI are inspired by, and are intended to, mimic the way that the human brain works. Neural networks consist of artificial neurons, and these artificial neurons, known as nodes or units, are connected in intricate patterns, which allow them to communicate and share information with one another.

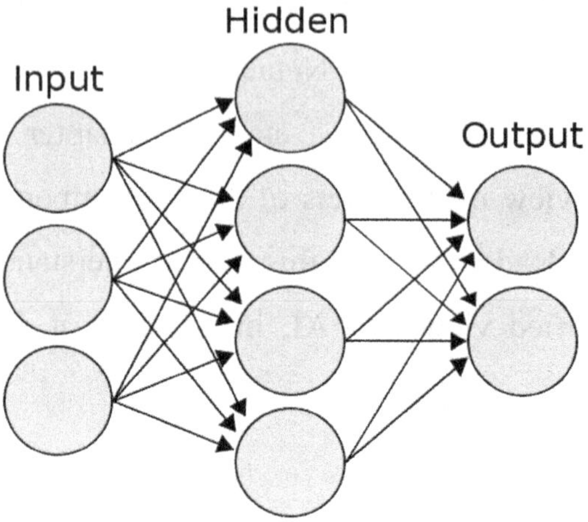

Figure 9: Neural Network (Credit: Free Software Foundation, published under Creative Commons Attribution-Share Alike 3.0 Unported license)

A simple Neural Network is depicted in Figure 9. As illustrated, imagine a series of interconnected circles, or nodes, arranged in layers. There is an input layer, a hidden layer (there can be multiple hidden layers), and an output layer. Each node represents an artificial neuron, and the connections between them represent the flow of information. In the input layer, the network is provided with some initial information, such as an image, a set of numbers, or any other form of data. This data is fed into the network, and each input node receives a specific piece of information. This information undergoes a transformation as it flows through the network. The hidden layers that are located between the input and output layers process the information in a series of steps, each layer building upon the previous one. These hidden layers help the network extract meaningful features and patterns from the input data. Finally, the processed information reaches the output layer, where the network produces its final result or prediction. The output layer consists of nodes that represent the possible outcomes or classifications based on the input data.

Each connection between the nodes has a weight associated with it. These weights determine the strength of the connections, and influence how information flows through the network. Initially, these weights are random or set to specific values. But as the network learns, it adjusts these weights to fine-tune its performance. This process of adjusting weights is called training. During training, the network is provided with labeled examples, where the correct outcomes are known. The network makes predictions based on the input data, and its predictions are compared to the actual outcomes. Through this comparison, the network calculates the errors or differences between its predictions and the correct outcomes. The network then uses an algorithm, like a teacher or a coach, to update the weights and reduce the errors. This is akin to the brain adjusting the strength of connections between neurons based on feedback and learning from mistakes. The network repeats this process for many examples, gradually improving its ability to make accurate predictions. This is how Neural Networks learn and improve in a nutshell.

Neural Networks in the Early 2020s

Neural Networks excel in a variety of applications across different fields. Neural Networks are being leveraged for image recognition since Neural Networks can recognize objects, faces, or even handwritten digits in images. Neural Networks analyze the features and patterns within images, enabling applications like facial recognition, autonomous vehicles, and quality control in manufacturing. Natural Language Processing (NLP), which shall be discussed in this chapter, relies on Neural Networks. Neural Networks are being utilized in language-related tasks, such as machine translation for translating languages, sentiment analysis, and chatbots. They help computers understand and generate human-like text, making communication between people and machines more natural and effective. Neural Networks are very handy in the Financial Services sector, particularly for conducting complex financial analyses, predicting stock market trends, detecting fraudulent transactions, and assessing credit risks, thereby aiding financial institutions in making informed decisions and mitigating risks. In the Healthcare Industry, Neural Networks are helping to diagnose diseases, analyze medical images like MRIs and X-rays, and in predicting patient outcomes, thereby assisting healthcare professionals in providing accurate diagnoses and personalized treatment plans.

As is the case with Machine Learning in general, there are important ethical considerations and challenges to consider with Neural Networks. Just as with ML, Neural Networks can potentially perpetuate biases that could be present in the data they are trained on. It is crucial to ensure fairness and mitigate biases to avoid unintended discriminatory outcomes. The transparency of Neural Networks is another challenge. They can be complex and operate as "black boxes," making it difficult to understand how they arrive at their decisions. The topic of "Explainable AI" (XAI), a concept to provide transparency and allow for humans to understand how an AI arrived at the decision that it did, shall be explored in this book. Neural Networks require vast amounts of data to train effectively. Therefore, as with ML in general, safeguarding personal information and ensuring compliance with privacy regulations are essential in an era of increasing data breaches and data privacy concerns.

As AI continues its exponential growth, so will the maturity and sophistication of Neural Networks. The combination of Neural Networks with other AI techniques, such as Reinforcement Learning offers many opportunities for the technology to advance rapidly. This combination allows computers to learn by trial and error, making decisions based on rewards and penalties. It opens up possibilities for autonomous robots, game-playing agents, and optimization in various domains. Another derivative of Neural Networks is the concept of Deep Learning. Deep Learning involves

Neural Networks with many hidden layers, and has seen many implementations in AI-related fields such as Computer Vision, Speech Recognition/Speech AI, and Natural Language Processing (NLP).

Understanding Deep Learning

Deep Learning is a derivative of Neural Networks and is intended to provide computers with the ability to learn and understand intricate concepts, with human-like capabilities. The goal of Deep Learning is to provide machines with the ability to mimic the complexity of the human brain, and dive into the depths of information to extract hidden insights.

At the heart of Deep Learning are Neural Networks with many hidden layers, hence the term "deep." Each layer in the network learns to extract specific features and representations from the input data, layer by layer, moving closer to a “deep” understanding. Deep Learning algorithms excel at recognizing patterns and extracting meaningful features from vast amounts of data. They can identify intricate patterns that may be imperceptible to people, enabling computers to make sense of complex information. It's comparable to being able to have computers find microscopic needles in a galaxy-sized haystack of data.

Deep Learning Networks have demonstrated extraordinary capabilities across various domains. They can accurately recognize objects in images, understand and generate human-like speech, translate languages, play complex games, and even assist in medical diagnoses. Their power lies in their capacity to learn and adapt to diverse tasks without relying on handcrafted features or explicit programming. The remarkable strength of Deep Learning lies in its ability to automatically learn hierarchical representations from raw data. Each layer in the network learns to extract more abstract and sophisticated features from the ones extracted in the previous layer. This progressive refinement allows Deep Learning models to capture complex relationships and make high-level interpretations.

Training Deep Learning models is a process of exposing them to large amounts of labeled data, where each example has an associated correct answer. The network learns by comparing its predictions with the correct answers and adjusting its internal parameters, called weights, accordingly. During training, the network starts with random weights, much like a blank canvas. It then processes the input data, and its predictions are compared to the correct answers. Through an optimization process, the network updates its weights to minimize the difference between its predictions and the correct answers. This iterative process is comparable to a Michelin-star chef continually updating quantities of ingredients for a dish until they achieve the "perfect" result they were seeking.

Deep Learning represents a monumental leap forward in the field of AI. It enables machines to learn from raw data, discover intricate patterns, and make intelligent decisions. By emulating the complexity of the human brain, Deep Learning opens doors to a world of endless possibilities. While Deep Learning has shown immense promise, it also presents significant challenges. Deep Learning Networks often require large amounts of data and computational resources for training. Ensuring the availability of high-quality, diverse datasets and the ability to efficiently train deep models remain important considerations. Interpreting the decisions made by Deep Learning Networks is another challenge. The complex nature of these models can make it difficult to understand why they produce specific outputs. Explainable AI (XAI), discussed later in this book, intends to help provide a framework in interpreting and explaining the decisions made by Deep Learning models, ensuring transparency and trustworthiness.

Chapter Seven – An AI Primer: Part 3

The chapters on demystifying AI are intended to underscore the fact that AI is not a monolith. Success with your AI programs will need to have best practices that are just as broad in terms of their applicability and utility to any one of the AI facets. As we close out our demystification of AI, this chapter will provide an overview of Speech AI, and Natural Language Processing (NLP).

Understanding Speech AI (TTS and STT)

The power of speech facilitates communications and interactions, allowing people to express thoughts, ideas, emotions, and exchange information. Effective speech carries with it vital context – concepts and nuances that are pertinent within the context of any particular conversation. The branch of AI known as Speech - Speech AI - seeks to replicate this form of effective dialog between computers and people. Speech AI can be broadly classified into two categories - Text-to-Speech (TTS) and Speech-to-Text (STT). The power of Speech AI is its ability to serve as an effective bridge in communications between people and computers, making it much more natural for humans to interact with computers, and specifically, with AI-powered systems. Speech AI makes communications with computers more accessible, approachable, and friendly.

Speech technology finds applications in various aspects of our daily lives. Exploring a few examples:

The premise of Text-to-Speech (TTS) is to enable computers to transform written words into spoken language. TTS takes written text - like a book, an article, or a message, and converts it into human-like speech. This technology is ubiquitous across mobile devices and other personal consumer electronics. Thanks to TTS, computers can read information aloud to humans, enhancing accessibility for people with visual impairments, or simply providing a more engaging way to consume content. TTS has matured to a level where TTS voices sound natural and expressive, versus monotonous and robotic when they were first introduced. TTS models are trained on extensive audio data of human voices, learning patterns, accents, dialects, tonality, and nuances of spoken language that only people could infer in the past. The advancements in TTS have allowed systems to produce speech that is almost indistinguishable from human speech.

As the converse implies, Speech-to-Text or STT, enables machines to understand human speech and convert it into written text. STT models are trained on vast amounts of audio data, teaching them to recognize and comprehend spoken words. STT allows us to communicate with our devices more naturally, making interactions with technology feel more human-like. STT is one of the most practical applications of AI in use today, popularized by its ability to allow for voice commands issued to consumer electronics such as smartphones.

Virtual assistants like Siri, Alexa, or Google Assistant use both TTS and STT to interact with us through speech, providing us with information, answering questions, and assisting with tasks. Speech AI adds value across a myriad of instances such as in voice commands, translations, transcriptions, audiobooks, and even automatically generating subtitles for content across streaming media.

As is a recurrent theme with any applications of any field of AI, Speech AI also presents ethical considerations that should be kept at the forefront of any Speech AI implementations. It is important to ensure that data is collected and used in a responsible manner. Speech AI should require implementations to secure appropriate consent from those interacting with the AI, and be diligent in explaining how data is collected, used, and stored in a way that a lay person can easily comprehend. As with every facet of AI, transparency is key to maintaining ethical deployments of Speech AI. The promise of Speech AI is to facilitate significantly easier interactions between humans and technology, making this communication user-friendly, natural, and intuitive. Transparency regarding data usage, voice data collection, and proactive governance to ensure biases and discrimination are avoided in Speech AI implementations is crucial.

Understanding Natural Language Processing (NLP)

Natural Language Processing, or NLP, is the facet of AI that focuses on having machines understand, generate, and interact with human language. Human language is the way we connect, learn from one another, and convey our thoughts and sentiments. Natural Language Processing aims to bridge the gap between human language and machine understanding, enabling computers to interpret and respond to our words and sentiments (via an AI field of study known as "Sentiment Analysis").

While language might seem effortless for humans, understanding it poses significant challenges for machines. Language is incredibly complex, and is filled with nuances, context, and variations. Words can have multiple meanings, and the order in which they are arranged can completely alter the intended message. Moreover, human language is rich with idioms, metaphors, and cultural references that add layers of complexity. NLP strives to teach machines the art of deciphering language and extracting meaning from text or speech. Just as we learn the meaning of words and phrases through exposure and experience, machines learn through exposure to vast amounts of language data. NLP works by having machines break down language into smaller units. These smaller units are known as “tokens”, and are usually words, or even smaller components of language, such as characters. They then analyze the relationships between these tokens, forming a structure that represents the meaning and context of the text, enabling machines to understand language in various ways.

NLP offers a plethora of invaluable capabilities that are in active use as of the first half of the 2020s. One such capability is “Text Classification.” Machines can categorize text into different classes or topics, such as determining whether an email is spam or legitimate, classifying news articles into different topics, or identifying sentiment in customer reviews. NLP is being extensively used for extracting information from text. Machines can extract specific information from text, such as identifying names, dates, locations, or other important details. This is useful for tasks like parsing resumes, extracting key facts from news articles, or analyzing medical records. NLP is also being leveraged to facilitate the translation of text from one language to another. Machine Translation systems, such as Google Translate, learn patterns and linguistic structures from large multilingual datasets to generate accurate translations, making communication across languages more accessible. A highly relatable use of NLP is in the ability of NLP to allow for machines to respond to human questions. With NLP, machines can understand and respond to questions posed by humans. By analyzing the input question, machines search for relevant information and provide concise answers. This capability powers virtual assistants like Apple's Siri on iOS devices, Amazon's Alexa, or website chatbots that assist users with queries. Another prominent use of NLP is for Sentiment Analysis, enabling machines to detect emotions or sentiments expressed in text. A widely deployed use of Sentiment Analysis is to help in determining whether comments on the internet, such as customer reviews, voice recordings of customer feedback, or social media posts are positive, negative, or neutral. This capability is valuable for understanding public opinion, analyzing customer feedback, or monitoring social media trends.

As with other aspects of AI as a whole, NLP is not without its growing pains and challenges. Some of these growing pains might be unique to the time of this writing. Given the vibrancy of AI as a field of study, it is highly likely that we would solve to reduce or eliminate these challenges in the next few years. That isn't to say that these issues might not be replaced with a whole new set of challenges. One of the challenges that NLP struggles with is ambiguity. Humans struggle with ambiguity too, but this problem is exacerbated with machines and their (current) inability to fully comprehend situations, and apply context and judgment as humans do. Language is inherently ambiguous, and words or phrases can have multiple interpretations. Disambiguating the intended meaning in different contexts can be a complex task for machines. Another challenge for NLP is that language varies across regions, cultures, and even individuals. Different dialects, accents, and linguistic nuances pose challenges for machines to comprehend and respond appropriately. Context is everything in communication and language, and language often relies on context to convey meaning. Understanding the context in which language is used is crucial for accurate interpretation. Machines must be trained to recognize and interpret contextual cues effectively. Finally, as with most implementations of AI, the responsible use of NLP technology is essential – especially given the power of language. Ensuring fairness, avoiding biases, and respecting privacy are critical aspects that warrant attention as NLP systems continue to evolve. The integration of NLP with other AI facets, like computer vision or robotics, will create machines that can understand and follow complex instructions, extract information from visual scenes, or engage in rich, meaningful conversations.

Chapter Eight – An AI Primer: Part 4

This chapter concludes the journey to demystify AI and provide an overview into several facets of AI. The intent to have a shared and non-technical understanding of the depth and breadth of AI is twofold. First, it is vital for business and technology leaders to know of these basic terminologies and what they mean at the highest, simplest possible level. To practice AI best practices, one must be AI literate – not necessarily fluent, but at the least, literate. Being AI literate will be a crucial skill to navigate your AI programs successfully. Second, a quick exposure into the depth and breadth of AI should inspire business and technology practitioners to realize how essential AI best practices are for your organization's success. This AI Primer concluding chapter will provide an overview of Computer Vision, and Robotics.

Understanding Computer Vision

What NLP is to language and speech, the aspect of AI known as Computer Vision is to sight. Computer Vision is a facet of AI that focuses on how machines can perceive, interpret, and understand the visual world. The power of sight allows humans to see and interpret the world. Humans with sight have the ability to recognize faces, navigate complex environments, and appreciate what they see. Computer Vision aims to replicate this human ability, enabling machines to see, understand, and interact with images and videos. Images are a rich source of information, capturing the visual essence of objects, scenes, and people. Computer Vision equips machines with the ability to comprehend the content and context of images, just as we humans do. By analyzing pixels and patterns, machines can extract meaningful insights and make sense of visual data.

The concept of Computer Vision is predicated on the analysis of pixels. Pixels are the minute building blocks that compose images. Machines break down images into individual pixels, which are akin to atoms of visual information. By examining the colors, shapes, and spatial relationships of these pixels, machines can discern objects, textures, and other visual attributes. One of the most fascinating capabilities of Computer Vision is object and facial recognition. With object recognition, machines can learn to identify and differentiate between different objects; being able to tell the difference whether an image is a car, a dog, or something inanimate such as a coffee mug. Similarly, with facial recognition, they can recognize faces, distinguishing between individuals and even detecting emotions displayed on those faces. Through algorithms and learning, machines can recognize patterns and features that define specific objects or faces. They analyze the contours, textures, and other visual cues to make accurate identifications, just as we humans do when we recognize familiar objects or people. Face recognition is at the heart of how most iPhone users unlock their mobile phones as of 2023, a technology found across several devices across multiple manufacturers.

Computer Vision goes beyond individual objects and faces. It also enables machines to understand the larger context and scenes depicted in images. Machines can recognize landscapes, indoor settings, or street scenes, comprehending the relationships between various objects and their spatial arrangement. This contextual understanding enhances their ability to interpret visual information accurately. Most mobile phone operating systems' photo features include these capabilities. In addition to being able to recognize and understand images, Computer Vision is also about extracting meaningful insights and information. Machines can analyze visual data to estimate measurements, detect anomalies, classify images into categories, or predict future outcomes.

There are several applications of Computer Vision across multiple sectors as of the first half of the 2020s. In the healthcare industry, Computer Vision aids in medical diagnostics by analyzing medical images like X-rays or MRIs, helping doctors detect diseases or abnormalities, analyzing medical images, and assisting in surgical procedures, enhancing patient care and outcomes. In the transportation sector, Computer Vision is enabling self-driving cars to perceive and understand the visual environment, interpret their surroundings, identify road signs, pedestrians, and other vehicles, and thereby ensuring safe and efficient navigation on roads. In the manufacturing sector, Computer Vision enhances quality control processes by inspecting products for defects, ensuring consistency and precision. In the entertainment industry, Computer Vision enhances the gaming and augmented reality experience by seamlessly merging virtual elements with the real world.

As this aspect of AI continues to evolve, Computer Vision will have to navigate some challenges along the path to full visual understanding. These challenges include the fact that varied lighting conditions, occlusions, and different viewpoints pose hurdles for machines to accurately interpret images.

Understanding Robotics

We conclude the journey to demystify AI where we started – the conflation of AI and robots. AI can exist independent of robots and robots can exist independent of AI (with some basic programming). Robotics in the first half of the third decade of the 21st century do little to inspire fear and trepidations that these robots will end up to be like the T-1000 from the Terminator series. Robots have long captured our imaginations, appearing in science fiction and tales of the future. Today, AI Robotics brings these visions to reality, creating machines that can perceive, think, and act in the physical world. Just as humans navigate their environment, manipulate objects, and accomplish tasks, robots enable us to extend our reach and enhance our capabilities. This vibrant field continues to make significant strides, with most robot manufacturers looking to introduce robots as personal assistants in homes, making them consumer electronics of sorts. Everything from the Roomba robotic vacuum cleaner made by iRobot, to robotic lawn mowers and snow clearing systems, the applications of smart – if not sentient robots – within homes, are endless. From clearing minefields, other military and defense capabilities, to exploring other celestial bodies within our solar system and beyond, humans have been successfully deploying robotics technology in the 21st century thus far. The AI facet of Robotics holds immense potential. Robots will become more capable, adaptable, and integrated into our society. They will continue to collaborate with humans, augment our abilities, and open new frontiers of exploration.

At the crux of Robotics lies the fusion of AI and the physical world around it. Robots are equipped with AI algorithms, giving them the ability to understand, reason, and make decisions. They can perceive the world through sensors, process information using AI techniques, and act upon their environment using mechanical bodies. To interact with the physical world, robots rely on perception - the ability to sense and understand their surroundings. They use various sensors, such as cameras, microphones, or touch sensors, to gather information about the environment. By perceiving their surroundings, robots can identify objects, detect obstacles, or even interpret human gestures. The world got its first taste of what robotics might look like in February of 2016 when Hanson Robotics introduced Sophia the Robot. Designed to be socially intelligent and aware of her surroundings, Sophia the Robot (the world's first non-human citizen and UN Innovation Champion) can converse with humans and respond to external stimuli as one would expect humans to. Just as has been the vision behind Sophia, AI Robotics are not intended to be purely machines. These robots possess a cognitive capability that allows them to think and reason. They have internal systems that are similar to a human brain, in that they process information gathered from perception. Despite their similarity in design, as of 2023, these internal systems are far from being as sophisticated as a human brain. Through the power of AI algorithms, robots can understand the meaning of the information, make decisions, and plan their actions. The physical embodiment of robots sets them apart from other AI systems. They have mechanical bodies designed to interact with the physical world and act upon their environment based on the decisions they make.

Robots can serve as personal assistants, performing tasks like housekeeping, companionship, or even helping people with disabilities to navigate their daily lives. Beyond the vision to make intelligent robots household companions, robots are deployed widely across many sectors as of 2023. In manufacturing, robots automate repetitive tasks in assembly lines, improving efficiency and precision. They can assemble products, perform quality inspections, and handle hazardous materials. In the healthcare sector, robots assist in surgeries, enabling precise and minimally invasive procedures. They can also aid in rehabilitation, providing support and assistance to patients. In agriculture, robots are used to automate agricultural tasks like planting, harvesting, and monitoring crops. They enhance efficiency and reduce labor-intensive processes. As discussed earlier, robots are invaluable for non-human exploration. Robots are used to venture into environments that are dangerous or inaccessible to humans. They explore other planets such as Mars (and are expected to travel to one of the Jovian moons within a decade), dive into the depths of oceans, or survey disaster-stricken areas. Collaborative robots, known as cobots, work alongside humans, sharing workspaces and responsibilities. They can perform repetitive or physically demanding tasks, freeing humans to focus on more creative or complex endeavors. This collaboration enhances human-machine interaction, creating a seamless fusion of intelligence and physicality. It's essential to establish guidelines and regulations that govern the development and use of robots to safeguard human well-being and ensure responsible deployment.

Chapter Nine – An AI Primer: Part 5 - Generative AI (A)

ChatGPT, the fastest growing consumer application in history, triggered an "AI Arms Race" in Q1 of 2023. Artificial General Intelligence (which includes ChatGPT, Large Language Models, Generative AI "GAI", etc.,) is a dramatically expanding new aspect of AI that will dominate and disrupt over the next several years.

Just a few months after introduction, not only did ChatGPT become the fastest growing application in human history, reaching 100 million users in less than two months, but it did due to its unique positioning as a tool that holds crossover utility across academia, personal lives, and professional settings alike. In contrast to this astonishing growth, the closest platform to reach 100 million users in record time was TikTok, taking 9 months to do so. To attain the same size of a user base took YouTube 1 ½ years, Instagram 2 ½ years, WhatsApp 3 ½ years, Facebook 4 ½ years, Twitter/X 5 years, Spotify 11 years, and 18 years for Netflix. As a fun contrast, the telephone was invented in 1878 and it took 75 years for the telephone to reach 100 million users. The mobile phone (not the smartphone) was invented in 1979 and took 16 years to attain the same user base. The internet started to become publicly popularized in 1990 and it took 7 years for the internet to reach 100 million users.

When we look back upon the start of the Age of AI, Generative AI's unprecedented and unrestrained growth in early 2023 will be remembered as the catalyst. An exploding field of AI, Generative AI are powerful algorithms that learn from vast amounts of data in the quest of giving machines a creative imagination, inspiring them to create art, music, stories, etc. Generative AI seeks to bring the human trait of creativity to machines, allowing machines to generate original and innovative content. Generative AI algorithms analyze patterns, styles, and structures in the data to understand how things are created and how to be creative. By learning these patterns, systems can generate new content that resembles the data they were trained on.

Understanding GPT and Large Language Models

Understanding ***ChatGPT*** requires a cursory understanding of ***GPT*** and the concept of ***Large Language Models***.

Similar to NLP (Natural Language Processing, as discussed in Chapter Seven – An AI Primer: Part 3), Large Language Models aim to bridge the gap between human language and machine understanding. Large Language Models are sophisticated AI systems designed to comprehend and generate human-like text. They are trained on vast amounts of data, such as books, articles, and websites, to understand the patterns and structures of language. By learning from this data, these models gain the ability to respond to questions, generate coherent text, and even engage in conversation. GPT-3 (Generative Pre-trained Transformer 3) is one of the most powerful Large Language Models. GPT-3 has learned from a vast range of text, absorbing knowledge from diverse sources. In interacting with GPT-3, humans can ask questions, give prompts, or seek information. GPT-3 analyzes the text provided and generates a response based on its understanding of language patterns and context, striving to respond in a way that sounds human-like, sharing information, insights, or even engaging in creative storytelling.

Large Language Models have the remarkable ability to generate text that is coherent, imaginative, and contextually relevant. They can write stories, compose poetry, draft essays, or even generate computer code. By learning from extensive text data, these models grasp the intricacies of language and use that knowledge to create new content. Large Language Models have revolutionized the way humans are able to access and interact with information. With their vast knowledge base, they can provide answers to questions, summarize complex texts, or even help navigate through the vast amount of information available on the internet. Large Language Models are poised to democratize information, making it more accessible, and bridging the gap between individuals and knowledge.

ChatGPT

Developed by OpenAI, ChatGPT, short for ***Chat Generative Pre-Trained Transformer*** is an artificial intelligence chatbot. ChatGPT is a large language model that uses artificial intelligence to hold text conversations with users. An AI model like ChatGPT is trained on books, articles, conversations, and other sources of text. It learns how humans express ideas, respond to questions, and engage in conversation with one another. What has made ChatGPT stand out is that it is an AI model that has been specifically trained to have conversations with humans, and have these conversations appear to be more "natural." ChatGPT has learned from a wide range of text and conversations, and therefore is knowledgeable across a range of different topics. When ChatGPT is asked a question or given a prompt, it looks for patterns and relevant information in its training data to generate a response. ChatGPT uses its understanding of language and context to provide meaningful and coherent answers. It tries to give responses that make sense and sound like they could come from a person.

DALL-E

DALL·E is an AI model that can create remarkable images based on text prompts. It has learned from a vast collection of images and descriptions, enabling it to generate unique visual content.

Similar to ChatGPT, this image generation platform learns from images and descriptions to create visual artwork. During training, the AI model examines diverse data and learns relationships between different elements, observing colors, shapes, structures, and the way things are put together. By getting a deep understanding of these relationships, the AI model can then generate its own unique content. When DALL·E is given a prompt such as "a dog playing the harmonica" it uses its understanding of visual elements and patterns to create an image that matches the description. DALL·E breaks down the prompt, analyzes its components, and uses its training to imagine what the scene being described might look like. It considers shapes, colors, and even emotions to generate a visually stunning image that captures the essence of the prompt.

Chapter Ten – An AI Primer: Part 6 - Generative AI (B)

The exponential growth of Generative AI is certainly a seminal moment in the field of AI itself, but it will more than likely prove itself to be a pivotal moment in how humans - and societies at large - learn, live, work, and play. The potential applications of Generative AI are vast and varied. Generative AI holds the promise to transform industries, streamline processes, alleviate burdensome rote operations, enable operational efficiencies, yield significant cost savings, all the while helping to enhance human creativity.

The Promise of Generative AI

The most common use of Generative AI is, as the name suggests, to generate new, original content. Platforms such as ChatGPT and DALL-E are generators of relatively limitless new content. Generative AI's use across industries is to help individuals across the value chain to generate content – from brainstorming and seeding new ideas, to drafting outlines for articles and blogs, to imagery and text content for marketing materials. Being able to leverage Generative AI as a brainstorming tool for idea generation, and then subsequent vetting and deliberation of those ideas by a team, can boost an organization's collective creativity and innovation. Platforms such as DALL-E are seen as a boon for marketing teams to develop unique imagery, often yielding precisely what is being asked for, without having to always worry about either developing new images or locating and then licensing them. From content marketing, social media management, to ad copywriting, Generative AI can create compelling, data-driven content and streamline marketing campaigns. The same principle holds true with designers of new products – be those products a new machine, or apparel. With a few text-based prompts, individuals can create product templates and ideas, greatly reducing the time the product designers and illustrators spend in drafting and crafting images as a fundamental part of the product development process.

Another element of Generative AI platforms is their ability to boost the value humans can derive from their interactions with AI. Generative AI can greatly enhance accessibility tools for individuals with disabilities, powering speech-to-text and text-to-speech applications, making information more accessible to a wider audience. Generative AI can help in content translations, as well as summarizing complex content into consumable, simpler versions, thereby not only allowing to ease language barriers, but also allowing for simplification of complex artifacts for consumption for a wider audience. Agnostic to industry, the inherent ability for Generative AI to interact with humans in a simpler, more engaging manner, presents an immense opportunity to Generative AI to enhance chatbots and digital virtual assistants. Infusing chatbots with

Generative AI capabilities allows for these systems to engage with customers in a more natural and context-aware manner.

Generative AI can provide an immense boost to workplace productivity by the automation of rote tasks. With the investment of $10B from Microsoft into OpenAI, imagine the value of Generative AI being an integral part of Microsoft Office products. This is the predicate behind the concept of Microsoft Copilot, an AI assistant that is integrated across the Microsoft suite of applications and services, including the Windows 11 operating system and the Office 365 productivity suite (Word, Excel, PowerPoint, etc.). Consider the amount of time organizations could save by having Copilot generating contextual documents in Word, or helping to author and summarize emails in Outlook. For instance, if a paralegal within a firm spends an hour drafting a legal brief, and if Copilot can be instructed to – with a few parameters – automate the drafting of this legal brief, even if Copilot can get the brief 80% of the way there, it would translate to 48 minutes of time savings. The function of most roles would then pivot from authors to becoming editors.

Within academia and the education sector, Generative AI can help in the development of personalized content and learning materials, generating interactive textbooks, and automating the creation of educational content. ChatGPT can act as a teaching assistant to students by answering questions and providing explanations – acting as a cross between a library, a personal digital assistant, and an internet search engine. Image generation tools such as DALL-E can create engaging visual content for a wide range of educational resources – from textbooks to teaching aides.

In what can be considered as a hybrid between brainstorming, product development, and content drafting, Generative AI is poised to add a significant boost to media and entertainment. Generative AI can not only be used for image creation for animated features or storyboarding of live-action film, it can also be leveraged to draft plotlines for movies and TV shows, generate dialog between characters, compose drafts of soundtracks, automate the selection and categorization of movie scripts and genres based on a studio's focus for a year (for example: 2 comedy movies, 2 romantic comedies, 2 action, 2 horror – and being able to rank the likeliest ones to succeed amongst the hundreds of script submissions), etc. While none of these will be devoid of the human element in its entirety, Generative AI opens up a new world of possibilities and unlocks opportunities for streamlining and process efficiencies. Some of these AI use cases within the entertainment industry (use of AI and digital recreation) led to the Screen Actors Guild-American Federation of Television and Radio Artists (SAG-AFTRA) having the longest strike in SAG-AFTRA history. The strike that started on July 14th, 2023, and ended on November 9th, 2023 resulted in the loss of an estimated 45,000 jobs and caused an estimated $6.5 billion loss to the economy of Southern California according to an article in Deadline (Deadline.com, 2023). This kind of disruption will undoubtedly manifest itself across many other industries as Generative AI use cases begin to cohabit with work traditionally performed by human subject matter experts.

A similar use case to vetting of submitted movie scripts can be applied in the medical field. Within the healthcare industry, Generative AI could be used to automate in the analysis of medical data and recommending possible diagnoses and prescribing treatments for a described set of conditions. The time taken to review, summarize, and extrapolate insights from medical journals can greatly help advance the medical field by enabling physicians to key in on new discoveries that lead to new medical treatments.

Within software development, developers spend a good amount of time in code reviews and optimization of code. This is intended to improve code and product quality and ensure that code is as defect-free as possible. While a fundamentally important facet of software development, it is also a time-consuming aspect of the process. Not only can Generative AI assist in authoring robust, functioning code snippets that can be modularized and leveraged in order to assemble functioning programs, it can also conduct code quality checks and identify potential areas where defects might originate. While it will not completely alleviate the need for human expertise, automating some of these repetitive tasks, and providing assistance to software development teams, will allow for IT teams to focus on how to best solve for the business problems and gain assistance from Generative AI for validating the efficacy of the solution itself. For cybersecurity, Generative AI can assist in threat detection, as well as in the analysis and generation of cybersecurity reports.

Generative AI Risks and Challenges

As with most implementations within the sprawling field of AI, we have no real precedent to learn from the risks and challenges posed by Generative AI. While there is no playbook, the risks and challenges of Generative AI outlined below draw from a pragmatic and practical approach on how firms are leveraging Generative AI within their organizations. And once again, as with most AI implementations, this will prove to be a highly fluid list. The risks and challenges outlined here as of the first half of the 3rd decade of the 21st century are highly elastic. Some challenges will be stymied by the end of the decade, while a whole new set of challenges should be expected to emerge.

While focused specifically on the use of ChatGPT, the challenges of Generative AI outlined below include, but are not limited to:

1. Corporate Email Addresses: Without education and enforcement, employees could be using their personal email addresses to sign up for ChatGPT, while others are using their corporate domain

email addresses. This is a risk on account of the fact that should ChatGPT be compromised, employee Personally Identifiable Information (PII) - which, in most industry definitions is a combination of First Name, Last Name, and Email Address - could be at risk.

2. The risk of using your corporate email addresses on a public site without vetting or review exposes your organization to what associates do online using your corporate domain.
3. The risk of using employee personal email addresses to interact about corporate-aligned work on a public site exposes your organization to having employees potentially intermingling their personal digital identities with your corporate intellectual property (IP).
4. Restricting or limiting access to ChatGPT via your corporate networks will do little to alleviate any IP concerns since employees could always expatriate your corporate data to the platform via their mobile devices. This opens up the risk that you might lose control of your digital assets on a public platform without any possible way to track or control them.
5. Employees have the ability to post proprietary, confidential, or company-specific information into ChatGPT. This includes all intellectual property, from code for code quality checks to snippets of research for ChatGPT to expand upon. There is nothing from an IT perspective that you can do to prevent individuals from doing this – it has to be policy and reinforced at

a local level. There are two concerns that dovetail from this risk: i. your IP being out in the cloud in the custody of someone else (GPT/OpenAI), ii. should GPT's data be exposed by a bad actor, your intellectual property will be exposed and compromised.

6. For employees who are entering your firm directly out of school, and are used to leveraging ChatGPT for their academic purposes, you run the risk of their presumption that the platform is acceptable to use in a similar manner in the workplace.
7. GPT (and others) have levels of service – a free version and paid option/s. Like with several other online products, you have limited governance controlling what employees are paying for online.
8. There does not seem to be widespread awareness that ChatGPT (as of Q4 2023) is a version that stopped learning in September of 2021. Therefore, the platform's information is slightly dated as of 2023.
9. There is a risk that your corporate content – including publicly available articles and citations – could appear in ChatGPT without attributions and citations. When intermingled with data from other data sources, there is a potential that your statements, positions, and thought pieces could be misconstrued or taken out of context. Using image generating tools such as DALL-E poses a similar challenge with your content from infographics to likeness of your employees.

10. Generative AI is still developing and continues to learn as it is enriched with good, pertinent data. However, there are situations where Generative AI models are inaccurate in their responses. Without subject matter knowledge of a particular topic, it can be challenging for humans to discern between fact and fabrication by the AI. These fabrications are known as hallucinations – occasions when Generative AI confidently engineers and asserts content that doesn't align with fact, or is nonsensical and does not align with reality. One such example is when Google's chatbot, Bard, stated an untrue claim about the James Webb Space Telescope. In response to a prompt asking about the James Webb Telescope, Bard responded that the James Webb Space Telescope took the very first pictures of an exoplanet outside this solar system. This information was false, and in fact, the first images of an exoplanet were taken in 2004, whereas the James Webb Space Telescope was not launched until 2021. Another example is when Meta's Galactica Generative AI product was asked to draft a paper about creating avatars (presumably for the Metaverse). Galactica, in drafting this paper, cited a fake paper from a very real author conducting research in a relevant area.
11. Generative AI platforms are designed to augment human intelligence, not supplant it. While there is significant benefit in adopting Generative AI for operational efficiencies, elimination of manual work, costs savings, etc., one can easily see how employees could be tempted to leverage Generative AI for many

more aspects of their work than originally intended. Although increased reliance on Generative AI might provide immediate, short-term value, there could develop an overdependence on AI, which would paradoxically stifle human creativity and stymie innovative thinking.

12. ChatGPT and other GAIs present a number of new and emerging legal issues (see Appendix A (Neuburger, 2023)).

KEY RECOMMENDATION ON GENERATIVE AI

Generative AI is poised to provide significant value (via productivity boosts, automation of rote tasks, inspiring creativity, being able to redeploy FTE, etc.), but also presents a new set of organizational challenges. The primary challenge with Generative AI as relates to enterprises is being able to establish and enforce corporate policies and guidelines that govern the safe and effective use of these technologies. Similar to the use of Google or Bing as a search engine, enterprises would benefit by not blocking the use of ChatGPT or restricting access. Companies do, however, need to establish guardrails and parameters for safe utilization.

Companies should establish governance around the safe and effective use of Generative AI within their firms. Generative AI offers several opportunities for companies, but also presents a new set of risks and challenges. These risks and challenges have necessitated the creation and ongoing maintenance of an AI Governance Corporate Policy document. An associated challenge is that this field is so vibrant and rapidly evolving that the pace at which these platforms are changing will require organizations to continually revisit these corporate policies and guidelines. It is therefore recommended that any corporate policy governing (or preventing) the use of Generative AI be revisited biannually. While most corporate policies in 2023 are focused on ChatGPT, they should implicitly consider applying these policies to ChatGPT competitors such as Bard. Continual updates to any corporate policy document should reference other emergent competitors, including but not limited to the subsequent iterations of ChatGPT itself, a platform that will continue to be significantly more intelligent and responsive than the preceding version.

SECTION THREE

AI BEST PRACTICES

SECTION THREE – AI BEST PRACTICES
Section Three – AI Best Practices is divided into seven interrelated parts:
PART A: AI CURRENT STATE explores the importance of extrapolating where in the AI journey your organization currently finds itself. Understanding your AI current state helps to assess which facets of the AI Framework© might require additional customizations for your organization's unique and specific needs.
PART B: THE AIM FRAMEWORK© presents the AI Framework©, which includes AI best practices as well as turnkey, yet extensible, tools that can be leveraged to ensure your organization derives sustained success from your AI programs.

AIM FRAMEWORK© SUPPORTING COMPONENTS
PART C: EXPLAINABLE AI (FAIRNESS AND TRANSPARENCY) takes a deeper look at the central concern across any facet of AI – ensuring that AI models are free of bias and proxy discrimination and presents an overview of Explainable AI.
PART D: DATA, DATA, DATA delves into what is at the foundation of all AI – data, and why it is of paramount importance to prioritize robust data strategy and governance programs that can provide clean, accurate, secure, and high-quality data for AI models.
PART E: LEADING AN AI-READY ORGANIZATION presents guidance on how each of us – leaders of people as well as practitioners – have to shepherd our organizations through the coming changes caused by AI, and champion a culture that thrives through change
PART F: "HUM-AI-N" – PERSONAL AI READINESS focuses on the paradox that to be most successful with AI implementations, we need to embrace traits, attributes, and qualities that separate us from AI – our innate HUMAN qualities and skills.

PART G: APPLYING AI BEST PRACTICES reviews some of the possible applications of these AI best practices across a myriad of industries and sectors in order to provide real examples of the best practices in action.

SECTION THREE

AI BEST PRACTICES

PART A – AI CURRENT STATE

Chapter Eleven – AI Current State Evaluation

We are on the cusp of an AI revolution that will impact every organization regardless of industry or sector. This AI revolution, the so-called "AI Era" or Age of AI, will swiftly usher in technological advancements that will usurp norms and disrupt entire industries, while giving rise to an untold number of industries and professions that we cannot simply envision today. Understanding and implementing AI best practices is not simply just a choice, but a necessity for organizations seeking to harness the full potential of AI while navigating the challenges that these unchartered waters present.

AI Program Maturity (2000-2030)

There has been a tangible difference in why companies across sectors have invested in AI programs within their organizations in the 21st century thus far. AI, as an ongoing evolution to organizational Digital Transformation programs, was transformative and proved to be transformational for some, but was disruptive for others. The COVID-19 pandemic only served to dramatically accelerate a slew of Digital Transformation initiatives that were already underway across industries and sectors. AI, as a natural evolution of technological maturity, was a beneficiary of this acceleration. Coming out of the pandemic, the "AI tsunami" washed over every industry and sector, whether they were ready for AI or not.

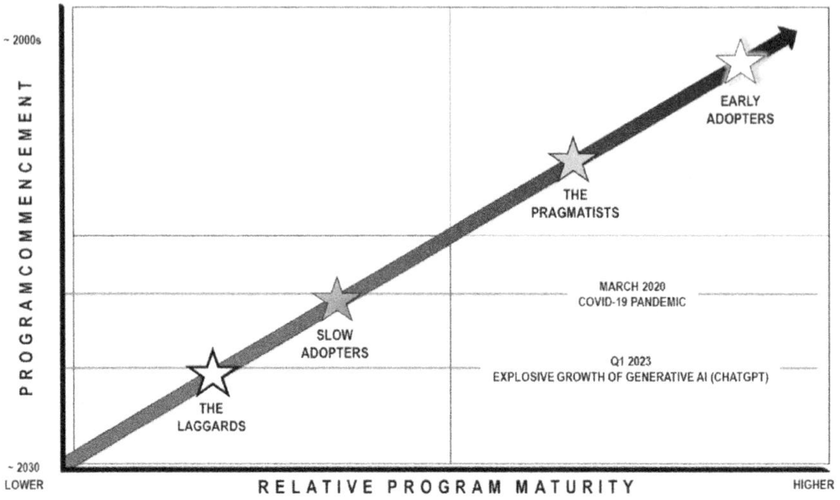

Figure 10: AI Program Maturity (in the 21st Century, 2000-2030)

As depicted in Figure 10, through the first three decades of the 21st century, one can identify four categories of companies as pertains to adoption of AI (and the consequent maturity of these AI programs) that have emerged across industries and sectors. These four categories are:

1. The "**early adopters**" - companies that had maturing data, analytics, and AI practices underway well before the pandemic.
2. The "**pragmatists**" - those who pragmatically invested in data, analytics, and AI practices before the pandemic forced them to, but not as early as the early adopters.

3. The "**slow adopters**" - those who were experimenting with data, analytics, and AI just before the pandemic and the explosive growth of Generative AI in 2023.
4. The "**laggards**" - those who were forced into action by the pandemic and the Generative AI explosion in 2023.

For the "early adopters" - companies that invested in data, analytics, and AI, starting in earnest at the turn of the century - their drivers had been customer experience, cost reductions, and process optimization. For these companies, their investments were about digitally transforming themselves by not just digitizing existing processes, but reimagining the processes themselves. Some companies within this "early adopter" group had been singularly focused on cost reductions as the primary driver as they sought to optimize and automate their rote, repeatable processes, in order to ensure that their operational tasks were optimized and automated. These companies saw AI and ML as decision-support systems and thought of these data-centric technologies as enabling augmented intelligence versus true artificial intelligence. Others that shared this "early adopter" group saw customer experience (CX) as an imperative for their organizations. Recognizing changing demographics, the need to appeal to a wider and younger customer base, and having invested significant sums into their digital transformations, these companies took a decidedly customer-focused approach towards their digitization, the scope of which included the entirety of their internal value chains. Organizations looked at Big Data / Decision Support Systems and rudimentary AI (including Predictive Analytics) as a way to improve their customer experience and save them money at the same time: a "win-win situation". The "early adopters" embraced data, analytics, and AI by forecasting where their industry was headed and by being encouraged that other "early adopters" in their industry or adjacent sectors of their industry were all headed in this direction.

The dichotomy between the firms considered to be "early adopters" to the "pragmatists" (those who had invested in data, analytics, and AI before the pandemic and Generative AI growth, but only after the "early adopters" had done so), was the "pragmatists" yielded to competitive pressures and the need to keep abreast with market demands as their impetus.

The "slow adopters" are the group of firms who were experimenting with data, analytics, and AI just before the pandemic, or had plans to commence their data, analytics, and AI journeys in the mid to latter half of 2020. The organizations who had some ongoing pilots or were on the cusp of operationalization of their programs had to scramble to institutionalize these practices. This scrambling was under intense pressure and duress for the potential impacts to their business and operating model during the pandemic lockdowns. The uncertainty during the pandemic's early days, when juxtaposed against some industries that are founded on certainty (such as Insurance), effectively proved to be quite disruptive for firms within these industries.

The "laggards" are the tiny fraction of firms that had pushed off data, analytics, and AI, and had no set plans in place to explore this practice. In this category, industries, and organizations within these industries, were slow to adopt digitization, or had done so with great disparity (digitized one facet of their operations and had not digitized the entire value chain). These organizations fared worse off heading into the pandemic in terms of their organizational readiness. To these organizations, the disruption resulted in a rush to find solutions. Companies that had been slow in their Digital/AI adoption or those that had not done so prior to the start of the pandemic, unsurprisingly turned to turnkey solutions that mature, established technology providers offered. The procurement of vendor solutions under relative duress also cleaved any given laggard industry into organizations that were larger/better funded and could rush into digitization by paying themselves out of the digital deficit they had found themselves in versus those who could not afford to do so.

For the firms who could afford to digitize in a compressed timeframe, under duress of the pandemic, the "build" versus "buy" choice for these firms always resulted in a "buy" decision. A "buy" decision did not warrant much other consideration since these companies did not have the luxury of time in order to build their AI/ML models and imbue them with value via external data sources. With the robust services that established technology providers deliver to companies across the ecosystem, it is unlikely that a company that has decided to go the “buy” route would choose to “insource” the work back into the firm and go the “build” route in the future. Most of these organizations would not choose to partner with a technology provider (“buy” their service) as an interim, stop-gap, while pursuing building out their own AI/ML models and using vendors/providers only for supplying external data. These firms might choose to build their own models in addition to, and not in lieu of, the services and value delivered by a technology provider.

Whether in the Finance or in the Healthcare sectors, the evolution of AI adoption across industries has been a journey marked by varying levels of sophistication and maturity. Understanding the historical context of those industries that are leading AI adoption, and those lagging behind, is crucial in comprehending the current positioning and future trajectories of AI within sectors. Understanding a company's positioning concerning AI adoption maturity within its industry serves as a general benchmark of where a company might currently be, and is fundamental to harnessing its potential effectively, by charting a course to where it wants to be in the future. Comprehending a company's positioning in terms of AI adoption maturity within its industry provides the foundation for strategic decision-making, risk assessment, fostering collaborations, and aligning with evolving customer needs. As AI continues to reshape industries, grasping its adoption landscape becomes increasingly essential for companies striving for sustainable growth and competitiveness. Despite being part art and part science, an understanding of your relative positioning will empower your company to navigate the complexities of AI implementation effectively, fostering innovation and driving success in an AI-driven world.

Your Industry's and Organization's Current AI Positioning

AI adoption maturity, and your company's consequent positioning related to this maturity, refers to the degree to which an organization has integrated AI technologies and strategies into its operations, services, and products. It is important to orient yourself to where your company might be in terms of AI adoption in advance of a study of the AIM Framework© and its constituent best practices. A current state assessment of your organization's overall AI posture is a partly qualitative exercise and unique to your organization. It is always helpful to understand your overall AI adoption maturity in order to surmise what aspects of the AIM Framework© your organization might need to invest more time in building out, or customizing to your needs. Your approach towards leveraging the AIM Framework© will need to be adjusted commensurate to whether your company already has a robust and relatively mature AI practice already in place, or whether you're starting from scratch. In any scenario - regardless of where on the AI maturity curve your organization resides - the AIM Framework© is designed to ensure long-term and sustained success with your AI programs. You can intuit your organization's AI maturity based on a myriad of factors, including but not limited to how technologically advanced the broader industry that your company operates within is, the gap between digital leaders within your sector and laggards, and the investments that your own firm has made in digitization. Your company's own AI maturity positioning is relative to AI maturity within your industry, and is inclusive of other companies within your industry.

As depicted in Figure 10, determining your organization's relative positioning, as mapped to the implementation of AI in the 21st century (through 2030), is part art and part science. Note that the AI adoption maturity landscape for companies within a specific industry, and between disparate industries can be quite broad and diverse. Some sectors, like Information Technology, and Finance (particularly for Stock/Portfolio Trading and Retail Banking), have been early adopters, leveraging AI for competitive advantage. There is a broad range of AI maturity that can be found even within any one specific sector. For instance, even within the Finance sector, while stock/portfolio trading and retail banking might be considered as early adopters, industries such as Property and Casualty Insurance could be considered as "pragmatists," while Life Insurance could be considered as "laggards." Even within one particular industry, you can find a broad disparity in AI adoption maturity. For instance, within the Life Insurance industry, AI and ML have been leveraged with rapidly increasing sophistication for the life insurance underwriting process; however, other parts of the life insurance value chain – billing, customer service, claims, etc. – might be comparatively immature.

Generally speaking, "early adopter" sectors such as IT, and some industries within the Finance sector, embraced rudimentary AI early in the 21st century. Some industries within the Finance sector established themselves as "early adopters" by leveraging algorithms for personalized recommendations, risk assessment, and algorithmic trading. Industries, such as Healthcare and Manufacturing, are progressively integrating AI across their value chain. Healthcare leverages AI for diagnostics and personalized medicine, while Manufacturing utilizes it for process optimization and predictive maintenance of machines. There are some industries that are still in the nascent stages of AI exploration that include industries such as Agriculture, Education, and some segments of Retail. These industries have been relatively slower in adopting AI (reflective of the fact that they were slower than other industries to embrace digitization).

Factors Influencing AI Adoption Across Industries

There are five main factors that have influenced AI adoption across industries leading into the 3rd decade of the 21st century, creating the four categories as depicted in Figure 10. These five factors can help you get an overall sense of where your specific organization is in terms of AI maturity within your broader industry and sector.

1. State of Digital and Data Sophistication:

Considered as a natural technological evolution, the adoption and maturity of AI within an industry, or an organization within an industry, is a direct reflection of the state of digitization and the sophistication of data practices within the industry or organization. Quite simply, the more digitally advanced or mature an organization is, the likelier it is to have adopted AI earlier than others. These organizations are often more sophisticated with how they manage their organizational data, and have robust data strategy and governance programs being actively practiced within their firms. Digital Transformations, which have been rampantly underway across every industry and every sector, are just as much about organizational culture and mindset as they are about tools, technology, and processes. Industries that have been further along in their digitization journey tend to foster a culture that has embraced innovation and data-driven decision-making. Those employed in organizations within these sectors often possess a higher level of digital and data literacy, and are more open to exploring AI applications. Highly digitalized sectors cultivate an innovation-oriented culture and nurture data-driven decision-making mindsets. This cultural readiness to experiment and incorporate AI-driven solutions facilitates faster AI adoption and integration.

Organizations with robust digital infrastructures often have better access to diverse datasets of high quality. The availability of structured and organized data allows for more effective AI model training and validation. For instance, sectors like Finance, which extensively collect transactional and user behavior data, can leverage this rich information to build predictive models and personalized services. Conversely, industries with lower digital maturity struggle due to data silos, poor quality, or inadequate data collection practices, impeding AI progress. However, this is not always the case. Some organizations might have advanced technological capabilities and systems, and could have significant investments in their digital transformations, cloud computing, etc., but could have immature data practices. These organizations are the ones who have traditionally thought of data as a by-product of their advanced systems, and not considered data as a core product. Having (mis)treated data, these companies find it challenging to pivot into investing in AI programs as having access to clean, and high-quality data is one of the key predicates to a successful AI implementation.

Organizations and industries with a higher level of digital maturity also typically have more advanced technological infrastructures. This equips the adoption of AI tools and platforms, and includes the utilization of cloud computing, high-performance computing, and scalable data storage solutions. As AI advances in some of the relative newcomers to the AI revolution, sectors like Healthcare or Manufacturing, which have adopted sophisticated equipment and Internet of Things (IoT) devices, can utilize AI for predictive maintenance, diagnostics, and process optimization.

2. ***Digital Disruptors and 21st Century Competition:***

Industries facing external digital disruption from digitally native competitors from outside their core industry often are challenged in continued investments in their digitization, and consequently are slower in implementing AI due to their focus on defending their core business. Examples of this include Airbnb disrupting the hotel industry, the taxicab industry being disrupted by companies like Uber and Lyft, Blockbuster being disrupted by Netflix, or retail titans such as K-Mart and Sears being disrupted by Amazon. Even within the same industry, organizations who failed in embracing 21st century digitization were imperiled and subsequently disrupted by competition (e.g., Nokia, as well as Blackberry, were disrupted by Apple's iPhone).

Industries that are primarily focused on battling external disruptions may lack the necessary expertise or organizational culture to effectively implement digitization, let alone execute on any AI programs. The emphasis on immediate challenges hinders the development of an AI-friendly environment within the organization. Industries under threat from external disruptors often exhibit resistance to change. This resistance can stem from a fear of further destabilization caused by adopting new technologies. The urgency to maintain market share and stability often overshadows the exploration of innovative AI solutions. Consequently, industries might overlook or delay AI adoption that could otherwise offer competitive advantages. When industries are embroiled in defending their core business against external disruptions, their focus remains on immediate survival strategies. This defensive, tactical stance can divert attention and resources away from investing in AI research, development, and infrastructure needed for innovation. The taxi industry, for instance, primarily concentrated on countering the disruptive business models of ride-sharing platforms rather than exploring AI-driven solutions for transportation optimization.

3. Infrastructure:

Industries dealing with vast datasets and complex algorithms, like IT and Finance, have had the ability to adopt AI earlier due to their existing technical infrastructure and capabilities. The elevated digital maturity of these industries has fostered infrastructural advancements that are conducive to AI integration, including sophisticated cloud architecture, agile high-performance computing, and scalable data warehousing. This factor, amongst the five, is most closely related to the first one. The distinction between this factor and the first one is that companies that are looking to commence their AI journeys can sometimes be discouraged from doing so – or delay their start – based on the upfront infrastructural requirements. Organizations that are mature in digitization can leverage their ongoing and existing digital investments, which renders AI adoption not quite as financially onerous in contrast to those just starting off.

*4. **Regulatory and Compliance Considerations and Constraints:***

Industries that operate within highly regulated frameworks or handle sensitive data can confront hurdles in AI adoption. Finance and Healthcare sectors navigate stringent compliance standards and privacy regulations, necessitating meticulous adherence to ethical AI practices. Despite having to ensure they are navigating the regulatory landscape, industries within the overall Finance sector have done relatively well in AI adoption. Having to balance innovation with regulatory compliance often elongates the integration process, compelling a cautious approach to AI implementation. In industries where compliance and security are paramount, the need to ensure that AI applications comply with regulatory frameworks can slow the adoption of AI, despite the industry's digital sophistication.

5. Talent Landscape:

Access to talent, both technology/AI and business expertise-related, influences the pace of AI adoption across industries. Sectors that are able to attract and retain talent tended to adopt AI more rapidly. There are industries within particular sectors that find attracting and retaining talent more challenging than others. Within the Finance sector for example, industries such as insurance have had a much greater challenge in attracting business and technology talent than industries such as institutional banking or investments. The ability to ensure a strong talent pipeline across business and technology is critical to a mature AI practice.

Why Extrapolate Your Organization's Current AI Positioning

While the AIM Framework© and AI best practices can be applied to any organization across any industry as is, it is worthwhile to extrapolate your firm's current AI positioning in order to understand what facets of the framework require extensions or customizations to be fit-for-purpose within your firm. This adoption maturity positioning encompasses various dimensions, including your company's technical capabilities, data readiness, organizational culture, and AI strategy. The next chapter will take a look at how you can extrapolate where your firm might be on the AI maturity spectrum, relative to your industry, providing qualitative and quantitative means of assessing your current state. In addition to being able to better apply the AIM Framework© to your unique needs, there are five reasons a current state assessment is a valuable time investment.

1. Facilitate Strategic Planning:

Equipped with the knowledge of your organization and industry's AI maturity can assist greatly in making data-driven, informed, strategic decisions. You can use your knowledge of your AI current state to identify opportunities for innovation, anticipate market shifts, and allocate resources effectively based on your industry's AI landscape. Risk management is a big part of an effective strategic plan, and understanding an industry's AI maturity helps in assessing risks associated with technology adoption. You are able to better anticipate challenges, such as data security or regulatory compliance, and prepare mitigation strategies accordingly. As AI continues its transformative journey, historical perspectives can guide industries and inform future strategies - in seizing opportunities, navigating challenges, and ultimately shaping a future where AI becomes an integral part of every sector's operations, driving efficiency, innovation, and growth. An exploration into relative positioning provides a roadmap for understanding the current landscape and for an exploration of future possibilities. "Early adopters" set the tone for innovation, while progressive sectors continually explore new AI applications. "Laggard" industries can learn from predecessors' experiences, facilitating smoother integration and fostering innovation.

2. Competitive Advantage:

By conducting an analysis of where you stand in relation to AI adoption within your industry will help in the identification of competitive gaps. Leveraging AI advancements strategically by using the AIM Framework© can then lead to differentiation and competitive advantage.

3. Ecosystem:

Organizations that are positioned at varying AI adoption stages within an industry can foster better collaborations and partnerships with third parties across their ecosystem. By being able to socialize, and potentially mutualize common problem solving, the entire industry might benefit, and AI adoption could be expedited.

4. Talent Acquisition, Development, and Retention:

An understanding of your industry's AI maturity, as well as that of your firm, aids in attracting and retaining top talent across business and technology. You can foreseeably tailor your talent strategies – from hiring to skilling and reskilling to align with industry standards and advancements.

5. Customer Engagement and Customer Expectations:

As with most things digital, industries with higher AI adoption maturity often shape customer expectations – on how customers expect to engage with a company. Understanding these expectations helps companies align their products and services accordingly to meet evolving demands within an industry.

Conducting An AI Current State Evaluation

The AIM Framework© introduces the concept of an AI Maturity Model (AIMM). Stylized to the Capabilities Maturity Model (CMM), the AIMM provides a five-stage ranking of AI implementation maturity within an organization. The AIMM allows organizations to benchmark their AI current state, infer the gap between their current state and the next level of maturity, and develop plans to tangibly mature their AI implementations from one stage to the next. For companies that attain the highest stage, Stage Five, the goal is to sustain this maturity level – sustaining success and maintaining a maturity level of Stage Five being a task that is more challenging than it appears. You know your organization the best, or at the least, can obtain pertinent information for you to conduct a current state assessment of your industry's and company's AI maturity. Some of this inference is going to be based on pure "gut feel." This is reflective of the relative newness of AI across sectors and industries, and even the disparity of how AI is being leveraged within a company across its value chain. Your assessment of the AI maturity within your firm - at this initial stage of developing best practices - can be relatively informal, and could be as straight-forward as applying those factors that are cited earlier in this chapter as a guide ("*Factors Affecting AI Adoption Across Industries*").

In order to provide some structure to this initial assessment, two question guides are provided below. Using the two question guides, you can conduct an initial AI evaluation of your industry's and your company's position to surmise where on the AI adoption spectrum your organization might currently be. This evaluation can be conducted in advance of leveraging the AIMM as a benchmarking tool (as part of the AIM Framework©) in order to continually improve your AI maturity.

Question Guide A – Understanding Your Industry's AI Maturity

The outcomes of this question guide provide companies with a base-level understanding of their industry's AI maturity compared to others. The insights you will collect, in combination with Question Guide B, will enable informed decision-making, strategic planning, and the formulation of tailored AI strategies aligned with the industry's unique demands and opportunities. This approach will allow you to better apply the AIM Framework© best practices, and position your organization at the forefront of AI adoption within your specific industry landscape.

INDUSTRY-LEVEL AI MATURITY ASSESSMENT

1. What is the state of digital play within our industry?
2. Have companies within our industry been embarked in a Digital Transformation, and if so, for how long?
3. How successful, by-and-large, has our industry been in effectuating digitization?
4. What is the general perception of our industry in terms of digital maturity?
5. How long have companies within our industry been leveraging data and analytics, including predictive and prescriptive analytics?
6. For how long has the term AI been used in the industry?
7. Are there known impactful AI use cases within our industry, and if so what are they?
8. Are there notable success stories showcasing significant benefits from AI adoption?
9. How are competitors leveraging AI technologies within our industry?
10. What are the latest AI trends and innovations specific to our industry?
11. Are there emerging AI use cases that could potentially disrupt or transform our industry?
12. What are the unique data challenges or complexities specific to our industry?
13. How are companies in our sector overcoming data-related obstacles for AI adoption?
14. How collaborative is the industry ecosystem in fostering AI innovation?
15. Are there partnerships or consortiums driving AI advancements within the sector?
16. What is the availability and demand for AI talent within our sector?
17. Are there notable skill gaps prevalent in the industry that impact AI implementation?
18. How do industry-specific regulations or compliance standards impact AI adoption?
19. Are there regulatory challenges hindering AI development within our industry?
20. How does our industry perceive and address ethical implications of AI applications, and what societal impacts or concerns are associated with AI use within our sector?

Figure 11: Guide A – Industry-Level AI Maturity Preliminary Assessment

Question Guide B – Understanding Your Company's AI Maturity

The answers resulting from this set of questions can provide an understanding of your company's current standing and capabilities in the AI landscape germane to your industry. In combination with your industry AI positioning as Question Guide A highlights, these insights will facilitate better application of the AIM Framework©, enable strategic decision-making, identify areas for improvement, and pave the way for a targeted roadmap for your organization towards enhancing AI maturity.

COMPANY-LEVEL AI MATURITY ASSESSMENT

1. What is the state of digital play within our company?
2. Has our company been engaged in a Digital Transformation, and if so, for how long?
3. How successful, by-and-large, has our company been in effectuating digitization?
4. What is the general perception of our company in terms of digital maturity?
5. How long have we been leveraging data and analytics, including predictive and prescriptive analytics?
6. For how long has the term AI been used in our company?
7. Are there known impactful AI use cases within our company, and if so what are they? Are these coordinated across the enterprise, or do we have disparate department-level AI activities underway?
8. Do we have an AI strategy? If so, how clearly defined is this strategy, and is it in alignment with our business objectives? Is this being appropriately communicated across the enterprise?
9. Are AI initiatives integrated into our overall strategic planning and corporate vision?
10. How adaptable is our current technology infrastructure to support AI? Are there limitations in scalability or compatibility with AI technologies?
11. What is the scope and impact of our current AI applications across various functions?
12. How are AI solutions contributing to business objectives, operational efficiency or revenue generation?
13. What is the state of maturity of our data strategy and governance program?
14. What is the quality of our data across various departments and systems? What is the level of organizational data literacy? How do we treat our data – as an asset or as a by-product?
15. Do we have access to all pertinent data? Are there challenges in accessing or integrating data?
16. Do we have a comprehensive inventory of AI-related skills among our workforce?
17. Do we have robust protocols ensuring ethical and responsible AI usage?
18. How effective are our AI governance policies in managing risks and compliance?
19. Are there potential risks or regulatory issues associated with our AI implementations?
20. Do we know how our customers might perceive and interact with AI-powered products or services?

Figure 12: Guide B - Company-Level AI Maturity Preliminary Assessment

SECTION THREE

AI BEST PRACTICES

PART B – THE AIM FRAMEWORK©

Chapter Twelve: The AIM Framework©

The AIM Framework© aims to serve as the underpinnings of a roadmap, guiding organizations - agnostic to industry and sector - towards successful AI adoption by outlining robust best practices tailored to drive value, foster innovation, allow for scalability, develop a culture for sustained AI success, and ensure ethical and responsible AI deployment. The AIM Framework® is intentionally scalable, extensible, and customizable. Organizations can leverage the framework and underlying tools in a relatively turnkey manner, or build upon the framework and make the implementation bespoke to an organization's unique needs. Regardless of the level of extensibility or customization, it is this implementation of the AIM Framework© and the underlying principles that will allow an organization to integrate AI into the very DNA of the organization.

As stated before, the field of AI, and AI's application has experienced explosive growth across sectors and industries in a short span of time. Although there have been industries and companies that are considered to be "early adopters" of AI/ML models and algorithms, these companies had little precedent to draw upon when establishing their practices. The less experienced firms in this space are scrambling to set up their own AI practices, under duress of the threat of competition, under pressure due to changing demographic expectations, and because the COVID-19 pandemic forced entire industries to rethink and reimagine digitization of their value chain.

While there has been a rush to embrace AI, even the earliest of "early adopters" have yet to step back and ensure that their talent strategies, their processes, and their technology are equipped to provide long-term sustained growth. Those slower to invest in these practices are still playing catch up, and are ceding foundational and basic aspects of establishing their AI programs for long-term success, trading them for short-term wins. All of this is occurring where there is keen public interest in AI, drawing global, national, and local attention from various legislative and regulatory bodies. This attention can be helpful since it will allow companies and technology providers to continue developing these programs with some parameters and guidelines that do not simply exist today. In this type of an environment, it will be crucial for companies with established programs to critically examine their own AI practices in order to ascertain that they are equipped for sustained growth, and implicitly continue to lead their industries, and for those who are still maturing on their journeys to set up their companies for success from the outset.

For mature industries with equitably mature and established practices, it is notoriously difficult to untangle and unwind processes and practices that become institutionalized within an organization.

Beset by high costs and afflicted with taking orders of magnitude longer than imagined it would take to do, decomposing existing industry practices and reconstituting them is a challenging task. This is a risk that any industry can ill-afford to take during shifting consumer expectations, changing regulatory landscapes, and the urgent need to stay abreast, if not ahead, of the AI curve. It is imperative for companies across industries to establish and adopt best practices that can serve as a de facto standard set of shared principles to build and develop their own programs from. The best practices and recommendations presented here are prescriptive in nature. They are not tailored to any specific firm, industry, sector, process, or technology. Intended to be as industry neutral as possible, these best practices are applicable to small and large companies alike, as they are to "early adopters" and "laggards."

The AIM Framework©

The AIM Framework© (illustrated in Figure 13) encompasses a set of twenty-five best practices. These twenty-five best practices are comprised of ten enterprise-level best practices, and five best practices each that are aligned to, and organized around, the "**People**," "**Process**," and "**Technology**" triad of building a successful AI practice. These twenty-five best practices include turnkey tools that organizations can implement as-is or can be extended and scaled to best fit your company's unique needs. The enterprise best practices are cocooned in four basic principles that organizations need to adhere to so that they can derive maximum benefit from the AIM Framework© AI Best Practices.

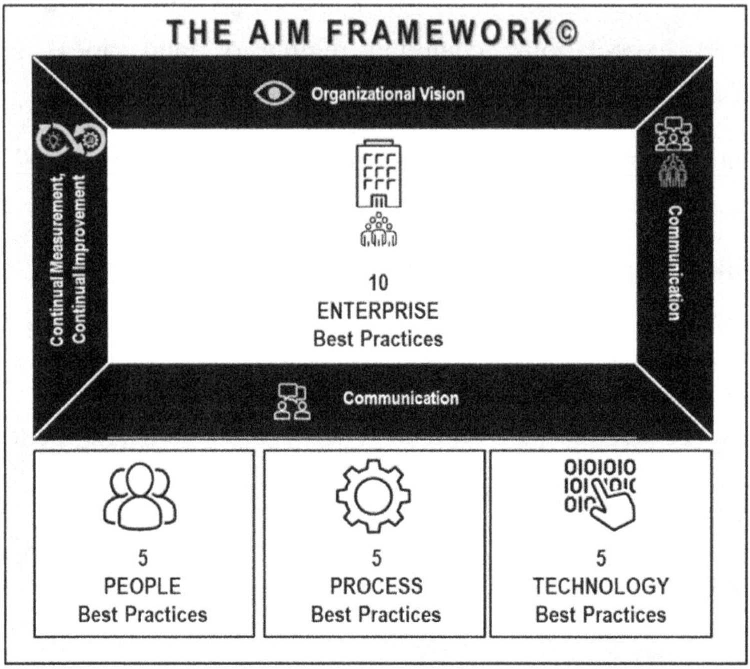

Figure 13: The AIM Framework©

At its core, the AIM Framework© is a standalone set of AI best practices. There are four supporting components in addition to this set of twenty-five best practices that organizations should consider in order to derive maximum value from their AI programs:

1. Explainable AI,
2. Data, Data, Data,
3. Leading an AI-Ready Organization, and
4. Personal AI-Readiness

Although you are free to consider applying each of the components in a standalone manner, the five components applied in aggregate – the best practices and the four supporting components – will greatly improve your odds of sustained AI success. Figure 14 depicts the AIM Framework© Best Practices as well as the four supporting components.

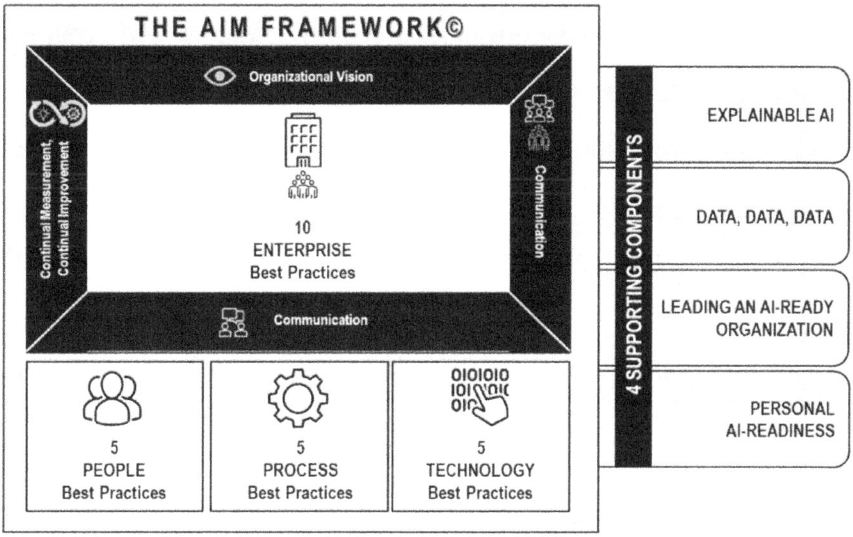

Figure 14: The AIM Framework© and Supporting Components

The People/Process/Technology (PPT) Framework

The AIM Framework© prescribes best practices that are aligned to the three core dimensions that are universal and form the bedrock to the success of any business in every industry – People, Process, and Technology. Based on being aligned to the familiar People/Process/Technology (PPT) Framework, the AIM

Framework© helps to identify how companies can organize their AI programs to achieve sustained success. The PPT Framework is highly popular across industries in order to maximize efficiency of their processes, ensuring the right tools are available for enabling organizational processes, and that the people enabling this work are able to seamlessly interact with these tools, operating within the bounds of these defined processes. The reason that these best practices are developed around the PPT Framework is simply because this framework is ubiquitous and commonplace across organizations regardless of industry. Firms look at most strategic programs through the lens of People, Process, and Technology, a practice that has yielded success over many decades. The fundamental reason behind organizing the AIM Framework© to the People, Process, and Technology triad is that sustained success will only come from a combination of ensuring the right processes are in place, run and championed by the right people, using the best technology suited for the task.

Depicted in Figure 15 below, the People/Process/Technology (PPT) Framework has been around since the early 1960s. First outlined by business management expert Harold Leavitt in his book entitled "Applied Organizational Change in Industry: Structural, Technological and Humanistic Approaches" (Leavitt, 1962), this framework has evolved since his first publication. There are several variations and depictions of the PPT Framework. Some represent the framework as a triad, or a pyramid, and others as a Venn diagram. The depiction of the PPT Framework represented in this book intentionally features "People" in the center of the PPT Framework.

In addition to the inherent familiarity that multiple industries have with the PPT Framework (in *practice*, not just in *theory*), it is also the bedrock of many-a Data Strategy and Governance programs across these industries. Organizations are in the habit of orienting programs such as Data Strategy and Governance, and Data Management around effectuating change in behavior and corporate culture (people), with processes and technology as facilitators and enablers respectively, around the PPT Framework. This is the reason why the "People" dimension features prominently in the center of the PPT triad presented in this book. The popularity of this framework, and its widespread use in data-related programs, is therefore best suited to structure these best practices in order to allow for the easiest way for companies to adopt these actionable recommendations, and implement them in practice.

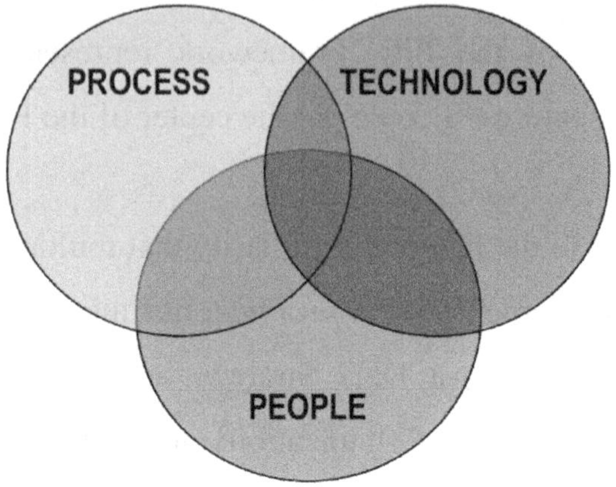

Figure 15: The People/Process/Technology (PPT) Framework

The following sections outline overall best practice recommendations at an Enterprise scale before delving into additional best practice recommendations around each, the People, Process, and Technology dimensions of the PPT Framework.

Enterprise Best Practice Recommendations

The AIM Framework© presents a set of ten enterprise-level recommendations for carriers to follow. While these recommendations are not meant to be an exhaustive list, they are intended to provide an elemental structural framework that all organizations can follow regardless of their maturity levels with their AI programs. These recommendations should be considered in addition to the ones outlined across the three pillars of the PPT Framework. Technology consulting firms, and technology providers should use these recommendations to guide, provide advice, and consult with their customers/partners. Some level of standardization across an industry based on these best practices, supported by companies, technology consulting firms, technology providers, etc., will allow for a stronger ecosystem holistically. ***Note that key findings and recommendations of this study are highlighted through this section and indicated as such***.

Figure 16 below depicts these ten best practices, along with four basic principles. These ten best practices should be applied at an enterprise level – that is, it is important that their application and adoption be across your entire company. Each of the ten enterprise-level best practices are mapped to domain/s within the PPT Framework – that is, each enterprise best practice can be ascribed to one or more of the domains of People, Process, or Technology. These four basic principles, navigating clockwise are "Organizational Vision," "Communication across the Organizational Hierarchy" (shortened to "Communication" on the vertical bar), "Communication across the Organizational Value Chain" (shortened to "Communication" on the horizontal bar), and "Continual Measurement, Continual Improvement." In Figure 16, they are represented as black shaded boxes surrounding the enterprise best practices recommendations framework, and are explained further below.

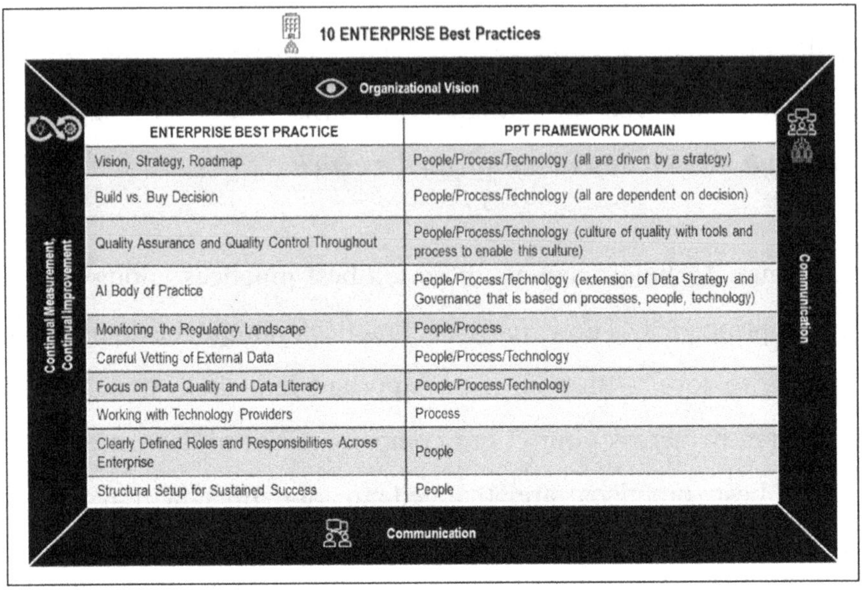

ENTERPRISE BEST PRACTICE	PPT FRAMEWORK DOMAIN
Vision, Strategy, Roadmap	People/Process/Technology (all are driven by a strategy)
Build vs. Buy Decision	People/Process/Technology (all are dependent on decision)
Quality Assurance and Quality Control Throughout	People/Process/Technology (culture of quality with tools and process to enable this culture)
AI Body of Practice	People/Process/Technology (extension of Data Strategy and Governance that is based on processes, people, technology)
Monitoring the Regulatory Landscape	People/Process
Careful Vetting of External Data	People/Process/Technology
Focus on Data Quality and Data Literacy	People/Process/Technology
Working with Technology Providers	Process
Clearly Defined Roles and Responsibilities Across Enterprise	People
Structural Setup for Sustained Success	People

Figure 16: Enterprise Best Practices Overview

People/Process/Technology Best Practice Recommendations

The AIM Framework© provides fifteen additional best practices (five each) across the People, Process, and Technology dimensions of the PPT Framework as highlighted in Figure 17.

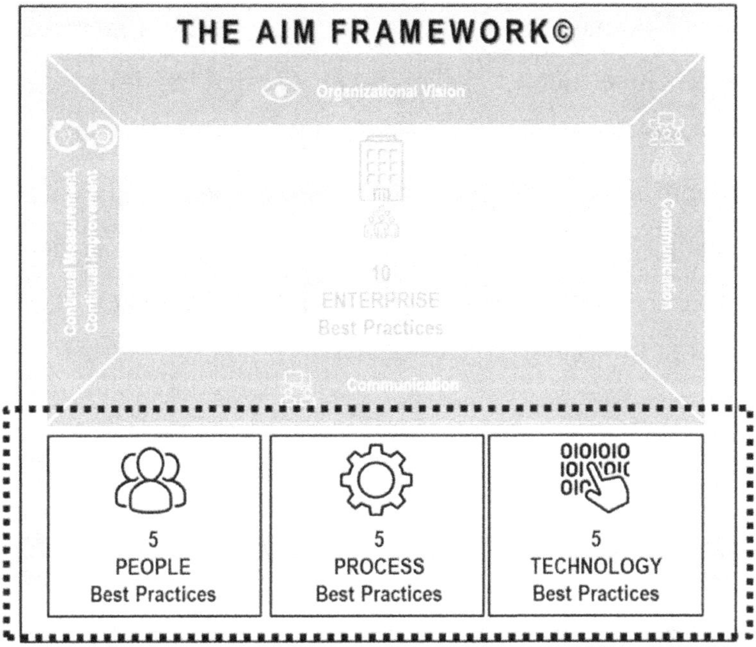

Figure 17: The Fifteen People/Process/Technology Aligned Best Practices

With the ten enterprise-level best practices serving as a foundation, these fifteen best practices provide a deeper look specifically organized around one of the three PPT domains. This categorization is a functional attribution to allow for an organization to consider who and/or what group should take the lead in championing a particular best practice. For instance, best practices aligned to the "People" domain require mindset and cultural changes, and therefore leaders – supported by their Human Resources groups – should consider taking the lead. Similarly, the best practices aligned to the "Technology" domain could be led by someone aligned to the IT organization within your firm, while the

"Process" domain changes can be led by the Enterprise Project Management Office, Agile or SAFe (Scaled Agile) practice leads, Operational Excellence, or simply individual/s who have a line of sight across your value chain and/or are enthusiastic about effectuating process changes/improvements. A tabular depiction of these fifteen PPT best practices, five within each domain, are depicted in Figure 18.

PEOPLE	PROCESS	TECHNOLOGY
1. Industry Domain Knowledge across Value Chain	1. Processes that Promote Transparency and Explainability	1. Standard IT Supported Technology Stack
2. Regulatory Knowledge across Value Chain	2. Customer Education on Use of Personal Data	2. 3rd Party Data Provider Taxonomy
3. Dedicated Teams and Resources	3. Documented, Repeatable Data Selection Processes	3. Automated Testing for Inadvertent Bias and Proxy Discrimination
4. AI and ML Governance Model ★	4. Documented, Repeatable Technology Provider Selection Processes	4. Automated QC Processes
5. Knowledge Sharing across Enterprise	5. Testing Rigor, Proactive Publication of Audit Results	5. APIs and Cloud as Core Requirements

Figure 18: The AIM Framework© - The People/Process/Technology Best Practices

A deeper dive into The AIM Framework©, starting with the Enterprise-Level Best Practices' Basic Principles, commences in the next chapter.

Chapter Thirteen: Enterprise Best Practices – Basic Principles: Part 1

The AIM Framework's© ten enterprise-level best practices are encased within four basic principles. These basic principles, as the name implies, are intended to help organizations to establish a solid foundation for the application of the ten enterprise AI best practices. The four basic principles (depicted in Figure 19) are:

1. Organizational Vision.
2. Communication across the Organizational Hierarchy.
3. Communication across the Organizational Value Chain.
4. Continual Measurement, Continual Improvement.

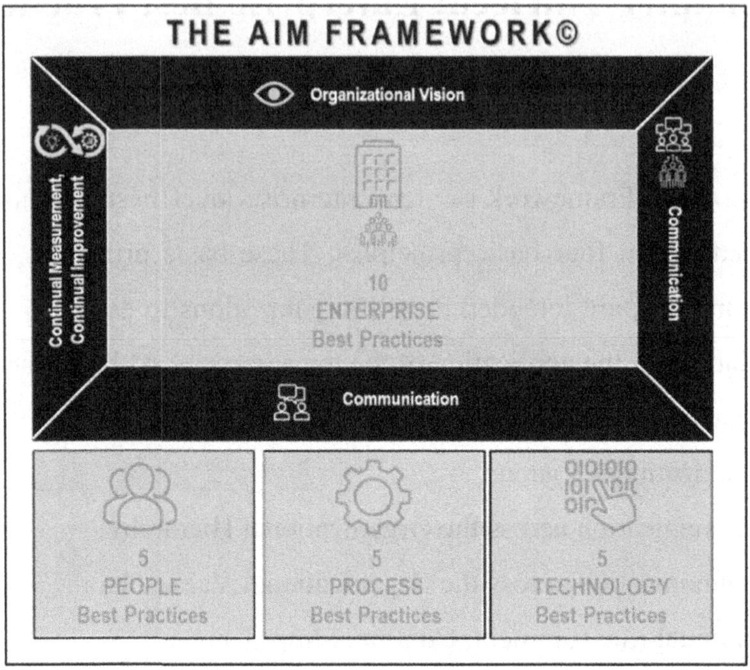

Figure 19: Enterprise Best Practices – Four Basic Principles

Basic Principle 1: Organizational Vision {⚡ KEY RECOMMENDATION}

AI, just as the case with every technology, is ultimately a tool. An exceptionally advanced tool that has the power to transform an organization, but a tool, nevertheless. Implicitly, how this tool is, can, and should be leveraged within an organization needs to be in direct support and furtherance of solving business problems. Some of these business problems can be those that an organization is encountering at the present, and others can be opportunities that AI as a tool can help forge for the company. Utilizing AI as a tool requires a direct and distinct correlation to an organization's vision and strategic plans. The application of AI within an organization therefore cannot be separate and distinct from an organization's enterprise vision and strategy. AI must be an inexorable enabler of your organizational vision. This will ensure the successful ongoing implementation of AI across your organizational value chain.

The application of AI as a tool in most organizations across industries has not been centrally coordinated at an enterprise level, falling instead to individual business units for experimentation, implementation, and advancement. This disparate adoption of AI disallows firms from having a coordinated, planned, and methodical implementation of AI. It prevents firms from being able to achieve scope and scale economies. "Organizational" AI programs have blossomed across companies more as autonomous principalities, rather than a federated or unified approach. The additional risk that this approach introduces is that while AI might end up being an indispensable tool for the successful achievement of individual departmental goals and objectives, the lack of central and interconnected coordination across the enterprise prohibits direct furtherance of a company's overall vision due to AI. A company would be unable to articulate the true holistic benefits to their business due to AI. A great risk with this scattershot implementation approach to AI is that an organization can truly never determine AI's value as a disruptor within their own company or across their industry. These firms would fail to recognize potential opportunities as well as anticipate potential competitors until it is too late for them to truly capitalize on their potential or mitigate competitive risks. This approach also presents a blind spot to an organization's ability to gauge skilling and reskilling opportunities within their enterprise.

The rapid pace of AI expansion - combined with the fact that these AI programs that operate in organizational silos - have not been a part of a core enterprise-level top-down objective. This could create misalignment between a company's enterprise vision and strategy versus what their AI program is evolving into. This could create misalignment between a company's enterprise vision and strategy versus what their AI program is evolving into. AI is going to continue to be a critical part of every facet of a company's operations. Implementation of an AI program within one facet of the organizational value chain is sure to exert influence onto every other facet of the company. It is therefore vital that these programs be a core part of an organization's vision, strategy, and roadmap. Without such an intentional approach to aligning AI objectives to enterprise objectives, it will be challenging for any company to marshal resources and appropriately organize themselves to the best practices outlined under the PPT Framework. Falling short in this intentional approach could mean the difference between equipping the enterprise for long-term, sustained success, versus having to invest in costly and disruptive rework of these programs in the case that they aren't effectively established.

Aligning AI with Enterprise Vision

The successful deployment and utilization of AI within your organization requires more than just technological expertise - it necessitates a strategic alignment with your corporate objectives and vision. The alignment of AI's implementation to an organization's vision must commence with the CEO, the organization's governing Board, and the C-suite on down across the entire organization. This basic tenet has less to do with a vision for AI and concerns itself more with alignment of enterprise goals with the establishment and execution of an AI program.

This basic principle recommends that these conversations of aligning AI with enterprise vision and strategy commence with the C-suite across companies. AI programs have taken root in enterprises via organic growth, having found themselves growing from ideation exercises, pilots, and proofs-of-concept, or at a focused-level as a partnership between business units and IT. While the C-suite are by-and-large aware of, and are enthusiastic supporters of these programs, there has been limited progress across companies to tie these programs specifically to their support of corporate goals, vision, and objectives. For instance, if a company has a corporate goal that states, "grow revenue," there is lack of clarity and alignment between the corporate objective and how AI is supporting this objective, despite clear evidence that ties this program to improved customer experience, reduction of operating costs, etc. There are a plethora of reasons that this alignment of AI with organizational vision and corporate strategy needs to commence with the CEO and the C-suite.

The expected value created by investing efforts into *Basic Principle 1: Organizational Vision* are outlined below:

Organizational Vision - Expected Value

Expected Value 1: Organizational Vision and Strategic Direction

Aligning AI implementation with corporate objectives and vision ensures that AI initiatives are purpose-driven and directly contribute to the company's strategic goals. When spearheaded by the CEO and the C-suite, this alignment sets a clear direction for innovation, emphasizing the 'why' behind AI integration, thereby preventing aimless technological adoption. Typically, the CEO and the C-suite, in concert with the Board of Directors or Board of Trustees in most organizations, is responsible for setting the organizational vision and/or aligning the company's strategic direction with the organizational vision. CEOs and C-suite executives possess the ability to allocate resources and prioritize initiatives across the organization. When they champion AI alignment with corporate objectives, it ensures that adequate resources, be it financial, technological, or people resources are allocated to support AI projects aligned with the company's strategic goals.

This group is also most responsible for ensuring that the corporate strategy is continually updated and reflective of changing market and business conditions. This group is usually the most influential within an organization and is responsible for lines of business and divisions within a company. They craft and execute strategy and are the seniormost leaders of the organization. The tone for AI needs to be set at the top and then cascaded throughout the organization.

Expected Value 2: Consistency and Standardization

Corporate objectives and vision serve as guiding principles for an organization's growth and success. Aligning AI initiatives from a C-suite level on down to these guiding principles ensures that AI advancements are consistent with the long-term strategic direction. This prevents a silo-based AI implementation – something that has been characteristic of most AI implementations across industries – and mitigates the inadvertent establishment of divergent paths that might hinder overall AI progress.

Expected Value 3: Effectuating Cultural Change

Within organizations, change, and particularly cultural change, always starts at the top. Endorsement of AI alignment from the CEO/C-suite down encourages cultural acceptance and adoption of AI initiatives throughout the organization. It sends an unequivocal message that incorporating AI into the company as a strategic priority is not just an IT initiative, but an integral part of the company's strategy. This helps to promote a cultural shift around AI, where employees look at AI as an inexorable part of the company's overall strategic priorities – as an enabler, and not just another project. This will foster a culture of innovation and adaptability across the organizational tiers.

Expected Value 4: Leadership

CEO and C-suite endorsement of AI alignment signifies leadership commitment and accountability. It sets an example for the rest of the organization, encouraging ownership and accountability for AI initiatives while emphasizing their significance in achieving the company's overarching vision. When the CEO and top leadership endorse AI alignment with corporate objectives, it fosters a collaborative environment. Departments and teams are encouraged to collaborate across functions, breaking silos, and working towards common strategic goals facilitated by AI. Aligning AI implementation with corporate objectives and vision is not about a technological strategy; it's a strategic imperative that requires top-level endorsement and commitment. When CEOs and C-suite executives lead this alignment, it permeates throughout the organization, guiding AI initiatives towards purposeful, strategic, and successful integration.

Expected Value 5: Corporate Governance, Stewardship, and Risk Management

CEOs and the C-suite are responsible to their stakeholders for being stewards of their companies by enacting appropriate governance structures for risk management and developing risk mitigation strategies. The shaping of an organization's ethical guidelines and the practice of adherence to these guidelines commences with this executive team. Aligning AI governance with corporate objectives ensures that ethical considerations and risk mitigation strategies are integrated into AI initiatives from the outset. This helps to minimize any potential adverse impacts on the company's reputation or operations that might arise from inadvertent ethical missteps or lapses in AI implementations.

"Corporate Allignment": Aligning AI with Enterprise Vision – A CEO/C-suite Guide

As we have established, the alignment of AI with organizational vision and corporate strategy needs to commence with the CEO and the C-suite. AI as a concept – and specifically one that can fundamentally transform their organizations and industries – can be somewhat of an esoteric concept to those executives that have yet to be immersed in AI for any meaningful length of time. It is beneficial for busy executives, who have limited mindshare, to be presented with a templatized starting point that they can use to facilitate dialog within the executive team to commence the alignment of AI as an enabler for their corporate goals. The twenty questions presented below should be used as a guide for executives seeking a starting point in being able to align AI program with their corporate objectives and vision. These twenty questions are termed as "***Corporate Allignment***", and should serve as a guide that executives can readily refer to as their organizations begin to craft/refine/review and execute on their AI strategies.

"Corporate Allignment" – Twenty Questions for the C-suite

1. What are your top five corporate goals over the next three to five years?
2. How does your AI program further these corporate goals?
3. How can you elevate your corporate AI program as a key strategic initiative, creating and communicating clear alignment between your five corporate goals over the next three to five years to your AI program?
4. Why are you investing in your AI program – is it to stave off competition, is it due to customer experience concerns, is it driven by cost-savings and efficiencies, or are you trying to

"Keep up with the Joneses"? These answers, starting with your "why," will drive how you can craft your program for the future.

5. Consider conducting an honest assessment of where your firm is compared to your industry as whole. Which category of AI adoption do you fall into? Is your firm an "early adopter" or a "laggard"?
6. What unique external pressures are you facing? Increased competition, changing regulation, digitally native startups, etc.?
7. What common external pressures are you facing? Changing socio-economic conditions, uncertain geo-political environments, shifting macro-economic outlook, disparate data privacy regulations, etc.
8. What internal pressures are you facing? Balance sheet concerns, stakeholders and investor concerns, legacy systems and technical debt, talent environment for hiring and retention, skilling and reskilling of staff, retention of corporate culture in a hybrid environment, etc.
9. How is your AI program of today equipped to help tackle or alleviate some of the external and internal pressures and stressors on your organization?
10. How mature is your firm's AI program? (Note that while question 5 helps you identify your AI adoption maturity, this may or may not be congruent with your organization's AI maturity. Your organization might have adopted AI earlier than

other firms within your industry, but the maturity of your AI program might still be developing as a practice).

11. If you believe that there is a palpable gap between where your firm is, and where the industry is as a whole in your AI practice, do you have a corporate goal to catch up?
12. If you believe that there is a palpable gap between where your firm is, and where the industry is as a whole in your AI practice, do you have defined plans on how you can catch up? (Note that having a goal and having specific plans actioned towards those goals can be very different things).
13. How mature is your technology stack, and can your technology sustain your corporate objectives as pertains to your AI journey?
14. How does your organization – from the C-suite/line of business heads to the entry-level employee feel about your AI program?
15. If you realize that there is a gap between your AI program's current state and your desired state, how much is the organization willing to invest in bridging this gap?
16. Have you as a firm clearly outlined the Return on Investment (ROI) and conducted a Cost-Benefit Analysis (CBA) with your AI program? Does this align to your corporate goals and is not treated as separate and distinct from the overall objectives?
17. How is your relationship with your technology providers – regardless of whether you build your own AI/ML models and consume data from these providers ("build model"), or are receiving a full service offering from them, wherein they

maintain the data as well as the AI/ML models ("buy model")? How close are you to their product roadmap and how does this comport with your AI plans and corporate objectives?

18. Do you believe you have a data-driven culture, and what is the executive team doing to nurture this culture?
19. Does your organization have a role at the helm of all your data-related programs (such as a Chief Data Officer or Chief Data Analytics Officer)? Does the enterprise have plans to create such a role if one does not currently exist? Do your AI and ML programs fall under this person's purview?
20. As pertains to AI, how close with the evolving regulatory and compliance landscape is your firm?

With an organizational vision that incorporates AI as a strategic enabler, the three subsequent chapters will delve into the three other Enterprise Best Practices as depicted in Figure 20, namely:

- Basic Principle 2: Communication across the Organizational Hierarchy.
- Basic Principle 3: Communication across the Organizational Value Chain.
- Basic Principle 4: Continual Measurement, Continual Improvement.

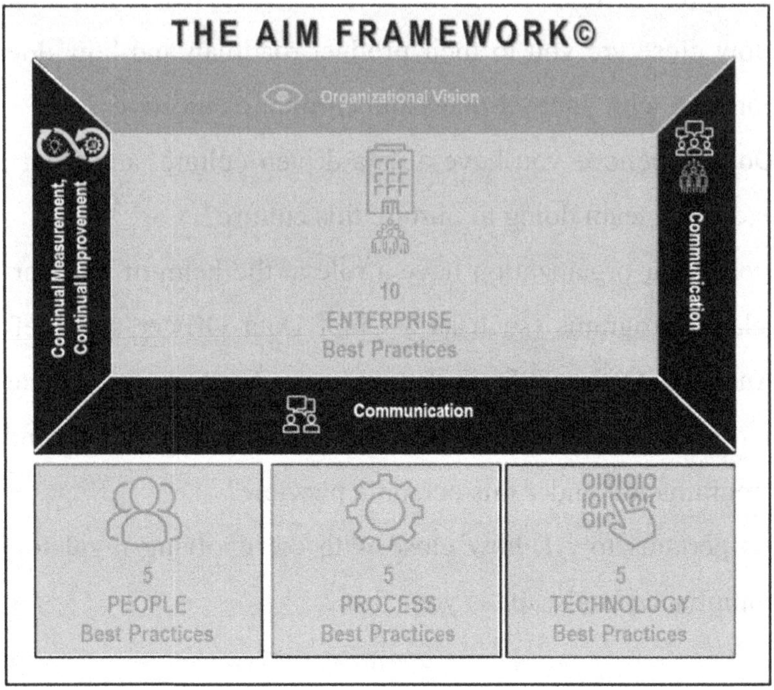

Figure 20: Enterprise Best Practices: Basic Principles 2, 3, and 4

Chapter Fourteen: Enterprise Best Practices – Basic Principles: Part 2

With the Basic Principle of Organizational Vision established, this chapter will explore Basic Principle 2: Communication across the Organizational Hierarchy as depicted in Figure 21 below:

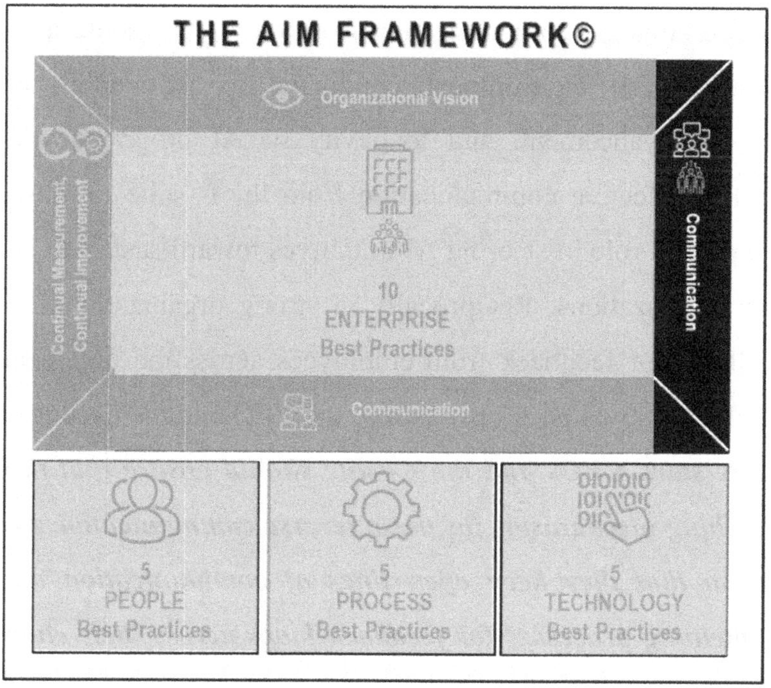

Figure 21: Basic Principle 2: Communication across the Organizational Hierarchy

ii. Basic Principle 2: Communication across the Organizational Hierarchy

Basic Principle 2 is focused on top-down communications across an organization. These top-down communications commence with the CEO and the C-suite, and are firmly understood, and then championed, repeated, and amplified by every tier of a company's leadership team (including all middle managers and team leaders). Clear and strategic communication from the CEO/C-suite is critical in fostering an environment conducive to innovation, agility, continual advancement, and longevity of AI programs within a company. Effective communication from the C-suite on down can play a pivotal role in steering AI initiatives toward sustained success within organizations. Reciprocally, a strong organizational culture also allows for feedback from employees across the organizational hierarchy, to travel back upstream to the CEO and the C-suite. ***From the very start, CEOs and the C-suite should ensure that they are establishing mechanisms for these reverse communication flows to occur so that they keep open lines of communication and are implementing processes for feedback loops across their enterprise tiers***.

The expected value created by investing efforts into *Basic Principle 2: Communication across the Organizational Hierarchy*, as well as how a company might get started, are outlined below:

Communications across the Organizational Hierarchy - Expected Value

Expected Value 1: Setting the Cultural Tone

Transparent, strategic, and inspiring communication establishes a framework for organizational alignment, cultural transformation, the marshaling of enterprise resources, and directs cohesive action toward realizing the potential of AI. This communication inspires the criticality of AI within employees, and sends a clear, unambiguous message that the firm is taking AI very seriously. In an era of constant and accelerating change that is driven by AI, communicating the vision, and continual reinforcement of the organization's approach towards AI, will ensure that employees' concerns regarding AI, and any job-related anxieties, can be proactively addressed. This top-down messaging will underscore the importance of ongoing learning, upskilling, and retraining – helping to relay your company's commitment to nurturing AI-related skills across your workforce. Communicating your organizational vision for AI-driven transformation will help to inspire confidence in your company's ability to adapt and thrive in an AI-centric future.

Top-down communications across the tiers of your organization will set the tone for your organizational enterprise AI strategy. Done correctly, this will establish an invaluable feedback loop that traverses the corporate vertical hierarchy. It is important that these communications should be easily understood by employees throughout the enterprise. CEOs and the C-suite should always remember to focus on the "*why*" behind each strategic decision, and keep revisiting this "*why*" as an anchor point. This "*why*" should outline organizational vision, goals, priorities, and provide a definition of what success might look like. Seeking to articulate the company's AI strategy transparently and easily, the communication needs to provide employees with your organization's vision and direction with respect to AI. Communications should seek to educate and motivate employees – underscoring the idea that each employee plays a key role in the company achieving their organizational goals. These communications are a vital tool for the C-suite to federate disparate AI implementations across your company, and to help aligning departments and employees across the enterprise. Clear communications will help establish a shared vision and common understanding of the significance and overall direction of your company's AI strategic objectives.

As is the case with most digital or data programs, effectuating a palpable change in your company's approach to AI will require a shift in organizational culture. It is strongly encouraged that the C-suite do not underestimate the power of communications from the top leadership to influence corporate cultural shifts that encourage an AI-focused mindset and fostering a culture of innovation and adaptability. Where it would be natural for employees to resist change and push back against change, some of this resistance being driven by self-preservation over the greater good, C-suite communication serves as a central tool in the promotion and advocacy for adopting AI across the value chain.

Expected Value 2: Creating Cascading Organizational Alignment

Executed well, communication and messaging can help significantly in driving enthusiasm and getting buy-in across the company. Equipped with a common understanding of your enterprise AI strategy, individual departments, including those that have active AI implementations, can be empowered to make informed decisions about their implementations at the local level. This will ensure congruence of these AI implementations to comport with the enterprise AI strategy, while nurturing innovation at the local level. Clearly tying individual departments goals and the objectives of employees within those departments to overall enterprise strategic goals will help to foster a sense of shared ownership, responsibility, and accountability. With these local implementations that align closely with the enterprise strategy, individual departments can socialize their lessons learned, learn from each other, and potentially mutualize the problem solving.

Expected Value 3: Operational Importance

On an operational level, having the CEO/C-suite provide guidance on AI as a priority, helps marshal resources across the company in an effective manner. This ensures that organizational investments align with the enterprise AI strategy. Having a central strategy that is well communicated throughout the company will prevent fragmentation and help in the development of a coordinated effort towards using AI in furtherance of enterprise goals. This allows for scope and scale economies to be realized. Applied to enterprise AI strategy communications, transparency around AI's opportunities as well as its impact on jobs and existing business processes allows for transparency and openness. This in turn can help significantly in building trust, which leads to greater confidence in the company's AI strategy. This will result in better alignment, adoption, and support for the strategy across your firm.

We have also explored that in the vast, sprawling industry that is AI, the primary concern across every branch of the AI tree continues to be explainability, fairness and transparency. Clear communications are necessary to ensure that employees understand that your organization's AI strategy reflects the highest standard of ethics that most company's core values espouse. These communications are your opportunity to establish ethical AI guidelines for your corporation. This will help provide clarity to your employees on your organization's perspective on adherence to responsible AI practices and how you plan to stay in compliance with evolving regulatory guidelines. The concept of Explainable AI (XAI) is discussed later in this book. Adopted in conjunction with the tenets of the AIM Framework©, XAI seeks to ensure that your enterprise AI implementations instill trust and confidence across the stakeholder spectrum.

Expected Value 4: Communicating the Criticality of Data

Finally, the successful deployment and scalability of AI is entirely predicated on data. This is true regardless of where AI is implemented within your organization's value chain. As part of a holistic enterprise AI strategy, these AI implementations within specific parts of your organization rely on data that is fit-for-purpose and fit-for-use. As established earlier in this book, AI programs will not succeed – or worse, provide flawed, erroneous predictions - without being based on robust, high-quality data. The way to successfully enable converting this data into meaningful information in an organization is by establishment of enterprise-wide data strategy and governance and data management/data quality programs. Therefore, it can be said that enterprise AI programs directly benefit from data strategy and governance and data management programs within an organization. Data strategy and governance programs are as much about effecting a cultural change throughout an organization as they are about processes or technology. Data governance is fundamentally about communication and influencing the entire organization via exercising this communication. Data management, and especially data quality programs, espouse the need for data literacy, and having a mindset around data quality be ingrained into an organization's culture.

The criticality of communication, especially top-down communication, has been repeatedly reinforced by several leading data scholars, experts, and practitioners. Thomas Redman, in discussing five steps to an information quality culture, states that "All change must eventually be top down" (Redman, n.d.). Data strategy and governance as well as data management programs can only attain sustained success by the vocal and visible support of the C-suite, specifically the Chief Executive Officer of an organization. Championed by the CEO as an organizational imperative, and amplified by the C-suite, the effectiveness of these programs throughout an enterprise is directly related to how critical employees perceive these programs to be. Treating AI as a corporate priority, by framing it as a supporting enterprise focus to corporate goals and objectives, in and of itself is only part of the story. It is crucial that the CEO and the C-suite communicate the importance of this program throughout the organization, continually repeating, reinforcing, and reiterating this message.

Getting Started

In concert with the C-suite, the CEO is responsible for articulating the overarching enterprise vision for AI integration, clearly explaining how AI, as an enabler, aligns with the company's strategic objectives. To quote a phrase that English-born American author and inspirational speaker, Simon Sinek, popularized - with a book of the same name - "Start with Why." Unambiguous and clear communications from the CEO, and amplified by the C-suite, about the "why" behind an enterprise AI program that seamlessly integrates into your corporate vision and objectives, helps employees understand the purpose and benefits. An educational mindset around the AI strategy not only helps in explaining – at an adequately high-level – the company's approach to AI, but can help demystify what AI means, and more specifically, what it means for your company and your employees. This will ensure employees refrain from succumbing to the fear of the unknown, and inaccurately extrapolate their own conclusions regarding AI's impact on your firm, the nature of their work, and their jobs.

The criticality of appropriate transparency when communicating down the corporate hierarchy regarding your AI strategy cannot be understated. Tonally, it is the CEO and the C-suite that nurture an atmosphere of open communication and transparency. As such, the CEO needs to ensure transparency in decision-making related to the enterprise AI strategy. Being able to transparently communicate -

and field employee questions and feedback - regarding the rationale behind strategic decisions, resource allocation, and prioritization of AI initiatives, will help to foster trust in employees. It is perfectly ok for the C-suite to acknowledge that the AI strategy is fluid and imperfect. Employees are often savvier than companies often anticipate in terms of their astute awareness of technological changes happening around them. Attempting to pass off an evolving AI strategy as immutable and perfected would likely be counterproductive to gaining employee's trust and their confidence in the C-suite's ability to have a firm grasp on their corporate AI plans.

The CEO and the C-suite must be willing to acknowledge that they might have blind spots, and not everything outlined in the AI strategy will yield desired results. The senior leadership team must demonstrate vulnerability that they might not have all the answers – and that is ok. Not having all the answers does not imply that no progress can be made. It is imperative that the C-suite clearly communicates that the company might not get everything right and that there will be setbacks and failures within the enterprise AI program. This is the C-suite's opportunity to foster innovation and experimentation, by dispelling employees' fears of failure. Creating a failure tolerant organization will be a key factor for your company's ability to continue innovating as AI continues to advance. It is more important now than ever for organizations to be able to fail fast, learn faster, and rapidly apply these learnings to another iteration. Enterprise agility will matter more in the Age of AI than ever before.

It is of vital importance that the CEO and the C-suite can effectively convey the AI strategy downwards across the company. Corporate communication needs to ensure that employees at all levels understand, embrace, and contribute to the successful alignment of AI implementation with corporate objectives and vision. At every level of the enterprise hierarchy, it is important that companies either piggyback on their existing feedback loop structures, or establish these mechanisms as part of the communication plan. This feedback loop can range from employees being able to provide feedback to their direct leaders and this feedback traveling upstream, or generic emails where employees can ask questions or provide feedback, internal Knowledge Management systems, internal community sites on the company intranet, Slack channels, etc. Most companies will typically employ an omni-channel approach – where feedback is provided directly because of employee team meetings with their direct leader, as well as a host of other digital feedback means, such as Knowledge Management, email, etc. Upper-level leadership, including the C-suite, must be diligent in ensuring that they're consciously and intentionally monitoring employee feedback. Leadership needs to ensure that they are acknowledging and/or addressing employee feedback in subsequent employee forums, reverting to the specific employee, and actioning pragmatic feedback, while recognizing/appreciating the employee/s who provided the feedback.

Once the C-suite has aligned AI, as an enabler, into the organizational strategic objectives, cascading these goals itself is not onerous. This is simply because most organizations already have ready mechanisms in place to develop organizational goals and cascade them throughout their enterprises as departmental goals, which in turn, are cascaded down to employees as part of their annual goals and objectives. The notable distinction here is that companies would do well to communicate that AI – as an enabler – is an inexorable part of these enterprise goals. Explaining to employees the rationale behind featuring AI as an enabler for corporate objectives, and how these objectives cascade down to departmental and employee goals, is important for employees across the company to clearly see that AI is an enterprise priority. Not all cascading AI-enabled goals are equal – there will be employees across the enterprise who will have little to nothing reflective of AI in their corporate goals. This is not to say that the jobs that these employees perform will be unimpacted by AI. Some employees will see the effects of an AI-enabled enterprise goal sooner than others would. It will all come down to a matter of prioritization of what initiatives the C-suite has decided to pursue for a given period. Companies would do well, as part of their communications, to be transparent with all employees that their companies – and potentially job functions – will change. The priority order of when these changes will be felt by individual departments and employees will vary based on what the C-suite agrees to is most pressing for any given objective setting cycle. Regardless of when employees directly see AI reflected in their personal goals, the fact that AI will be integrated into the enterprise strategy will underscore that AI is poised to directionally be a core part of your organizational goals.

For several companies, these kinds of top-down communications will be finessed and managed by the Corporate Communications/Internal Marketing teams. Once the initial communications conclude, the CEO can then provide regular updates on the enterprise AI strategy implementation progress, and its alignment with corporate goals. These regular communications should be tailored to the employee base, as well as other stakeholders, such as the Board of Directors. The CEO and the C-suite can continue keeping employees apprised of progress – including wins, challenges, upcoming risks, and mitigation plans, expected opportunities, etc. – through a variety of forums. These include all-employee events, town halls, small group meetings, CEO emails, periodic newsletters, etc. It will be important for any employee-related communications to highlight the contributions of some employees working on AI-related initiatives, and the criticality of all employees in achieving AI-related objectives.

When to Start

Do not let "perfect" be the enemy of the good. Given how vibrant the field of AI is, the rapid pace of AI growth will make it very challenging to have a "perfect" enterprise AI strategy. It will be an error to pursue any objective definition of perfection for the enterprise strategy before corporate communications occur. Therefore, the enterprise AI strategy, outlining how AI will serve as an enabler to further corporate goals and objectives, need not be "perfect" before the C-suite commences articulating the organization's evolving AI strategy. Given the pace of change, it will be highly unlikely that any AI strategy can truly be considered as "done." Most employees will be perfectly comfortable with knowledge that an evolving AI strategy is being deployed across the enterprise, and that their leaders have a cogent plan to capitalize on AI's opportunities and mitigate the risks posed by AI. It is strongly encouraged that organizations do not wait to attain a nebulous definition of perfection for their AI strategies in order to communicate, and certainly not wait to develop their communication plans until after this "perfection" has been achieved. Communications from the C-suite on down that emphasize the AI is rapidly evolving and your firm stands ready to evolve with it, emphasizes adaptability, and fosters agility in responding to AI advancements and changing market conditions. It is also highly recommended that organizations plan to ensure that these communications are not “one and done”. These communications must be featured regularly and prominently as an integral part of regularly scheduled employee communications, from CEO newsletters to “all hands” meetings and employee town hall type forums.

Chapter Fifteen: Enterprise Best Practices – Basic Principles: Part 3

With the Basic Principle of Communication across the Organizational Hierarchy reviewed in the previous chapter, this chapter will explore Basic Principle 3: Communication across the Value Chain as depicted in Figure 22 below:

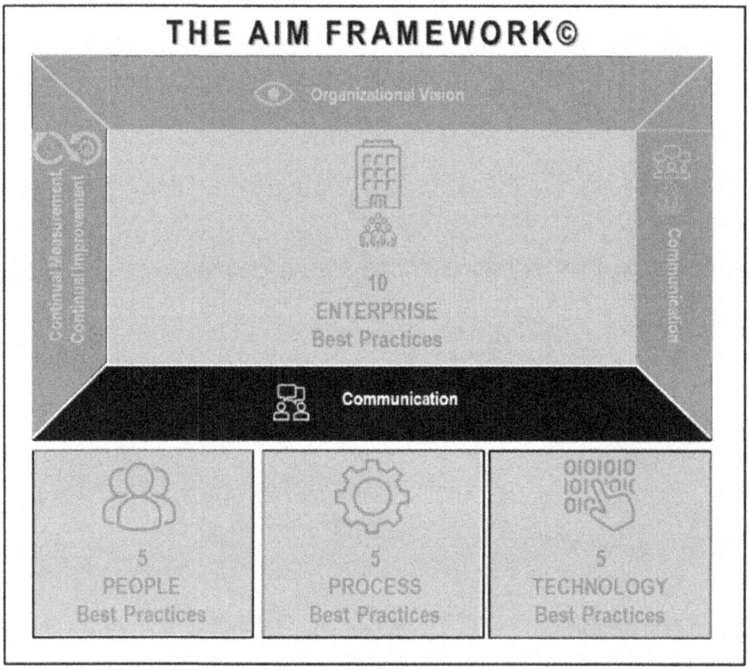

Figure 22: Basic Principle 3: Communication across the Value Chain

iii. Basic Principle 3: Communication across the Organizational Value Chain

Just as communication is critical across every tier of the corporate hierarchy, especially when it comes to treating AI as an enterprise priority and attaching the program to corporate objectives, it is just as crucial across the organizational value chain. This is depicted as a horizontal bar that traverses across a firm in Figure 20. In addition to top-down communication from the CEO/C-suite across the enterprise, the success of AI programs is greatly contingent on the ability of an organization to foster robust communication and collaboration among its various departments. Building a culture of mutual interest, transparency, openness, knowledge-sharing, and aligned objectives across the enterprise sets the stage for sustained AI success. Departments can inherit their own AI-related objectives from the broader enterprise strategic objectives, coordinating with their peers to achieve economies of scope and scale. Acting in coordinated unison, this cohesion helps in driving innovation, realizing operational efficiencies, and reducing costs in the rapidly evolving AI landscape.

The expected value created by investing efforts into *Basic Principle 3: Communication across the Organizational Value Chain*, as well as how a company might get started, are outlined below:

Communications across the Organizational Value Chain - Expected Value

Expected Value 1: Breaking Through Organizational Silos

Horizontal communication underscores the importance of having divisions/departments across the enterprise share equitably in their shared understanding of enterprise AI vision and strategy, and how much and what their contribution is to this holistic strategy. Ensuring horizontal communication channels are always open ensures that there is clear alignment within and across an organization to the overall goal and strategy. The central predicate of Agile, and Agile Scrum as a software development methodology is founded on cross-functional teams across the value chain coming together with a shared set of goals and objectives. Where an enterprise AI practice can be central to a firm, and touches upon multiple facets of the organization, it is important to draw from the Scrum concepts, and disperse this spirit through the enterprise. Agile methodologies can help in enabling rapid adaptation to changing AI trends or business circumstances. The open, honest, and transparent communication that Agile evangelizes will help to foster organizational adaptability, flexibility, and responsiveness to changing conditions.

Effective communication across the enterprise is a vital antidote towards eradicating the artificial silos that most organizations establish and institutionalize within their companies. Whether intentional or organic, silos within a company, regardless of sector or industry, are common. Based upon department or function – such as IT, Sales, Marketing, Finance, Human Resources, Product Development, etc. – and often then associated with bespoke technology platforms and institutionalized processes, the boundaries between these departments only tend to harden over time. This creates significant friction within an enterprise during execution of corporate initiatives or enterprise-wide systems implementations. Each department operates within their own parameters and processes, and somehow these disjointed processes get cobbled together to collectively execute on shared opportunities. This makes every enterprise-level endeavor a lot more complex and wrought with inefficiencies than it should be. This silo-based functioning prohibits operational efficiency and therefore tamps down the opportunity for the firm to achieve a modicum of maturity in any practice. With broadening operating gaps between departments, communication across the enterprise tends to suffer, which in turn broadens these gaps even more. Establishment of an enterprise AI program that is in support of corporate goals and objectives, and well communicated from the C-suite on down, cannot produce lasting results without cross-enterprise communication.

Digital transformations across several industries have thematically followed this pattern of inconsistent digitization across an enterprise. These industries have invested significant sums into digitization, but this digitization seems to have been clustered around the traditional silos of a firm. In the first quarter of the 21st century, companies that have invested in digitization have largely focused their transformations on digitization based on functional areas of their companies. Processes, platforms (technologies), and people have been aligned around these areas. Across companies, these functional areas are usually existing department-based silos; for example – functional areas such as Marketing, Sales, IT, etc., each operating in a silo. Each departmental silo has its own processes, which are then tightly intertwined around systems and their specific inadequacies. A typical scenario is that an organization has invested in modernizing one department's functions and systems within the company, but the systems of a different department within the same company are not up to par. Often, these companies modernize one aspect of their firm, but many other aspects continue to operate with disparate and distinct processes, outdated legacy systems, etc. These types of digital transformations can be referred to as the "peaks and valleys" approach, or a "digital rollercoaster ride."

Workarounds for system inadequacies have also been institutionalized, exacerbating the problem of systems that "don't talk to each other," and are a challenge to integrate. This manner of digitization has led to something known as the "Stack and Silo Effect," where digitization efforts have taken place within departmental silos, stacking existing platforms within department silos with additional digital platforms. It is critical to upend traditional ways and thought processes around how to digitize. It is inadequate to digitize within each silo. Vertical digitization will lead to newer and more current platforms, yet it will fall short of providing a smooth and seamless digital experience to its customers and stakeholders along the value chain. Applied to AI across the enterprise, a company would do well to keep their customer as their true north and digitize horizontally across the digital value chain. This ensures that there is a shared sense of purpose throughout the organization and distinct departments are invested in contributing to support the overall enterprise strategies as pertains to AI.

It is strongly recommended that companies pay special heed to their organizational silos. A siloed mindset will prove detrimental and hinder your company's chances of success with AI. Transparent communication between departments will help to dismantle silos, fostering a culture of transparency, information-sharing, and mutualizing problem-solving.

Expected Value 2: A Holistic Approach to Planning and Execution

An enterprise strategy that is not followed consistently across the enterprise is not an enterprise strategy, and will not yield the desired results. Given how rapidly AI is transforming industries, organizations cannot afford to lose time and squander opportunities to make progress because of miscommunications across the enterprise. Clear communication horizontally across the enterprise will help to align enterprise AI initiatives with overarching organizational goals, and cascade them down to departmental goals and objectives. This will thereby mitigate conflicting departmental priorities, or duplicative and/or redundant efforts. Departments can then synchronize their departmental efforts to prioritize AI projects that will support enterprise goals to maximize derived organizational value. This strategic coordination across the value chain will also help to prevent "AI cottage industries" from mushrooming across the company. These "cottage industries" can detract from the overall enterprise AI strategy by having their own agendas and priorities that might not fully comport, or be misaligned with, those of the enterprise.

Even without the presence of these "AI cottage industries" scattered throughout the enterprise, the value of an enterprise AI strategy resides in the ability to socialize knowledge and problem solve in a collective manner. Having each function within a company pursue their own versions of AI strategies (with good intent) is inefficient and wasteful. Communication across the value chain can help with the optimal utilization of organizational resources by helping make resource allocation more effective. In addition to avoiding potentially duplicative investments, communication allows for leveraging shared resources across the enterprise (or contend for the same pool of resources, which is where horizontal AND vertical communication conduits are so critical to help the C-suite prioritize AI initiatives at an enterprise level). This feedback mechanism can also help nurture learning across departments, enabling them to leverage insights and lessons learned from AI projects in other parts of the company. Being able to prioritize at an enterprise level allows the C-suite to reduce friction in AI programs by helping to centrally manage program and project governance.

Expected Value 3: A Holistic Approach to Enterprise Intelligence

A cross-functional approach that supports execution of the enterprise AI strategy, predicated on horizontal communication, allows for coalescing a diverse set of perspectives and experiences from across the company. Bringing together domain-specific subject matter expertise, business and technical acumen from across the firm builds enterprise AI solutions that are much more inclusive. This inclusivity ensures that the enterprise AI strategy does not lopsidedly favor one part of the company, while causing unintended consequences on a different part of the organization. Siloed environments are notorious for hiding risks as well as opportunities that exist within a company. This horizontal collaborative communication will facilitate identification of various use cases across the value chain that might not otherwise be visible.

Expected Value 4: A Holistic Approach to Enterprise Data Assets

Most organizations, even those that inhabit technologically progressive industries, struggle with managing their data on an enterprise scale. As soon as a data asset enters an organization, it tends to be duplicated and dispersed across the ecosystem. With limited - or siloed - data infrastructure in place, companies can never fully gain a handle on data asset provenance and taxonomy. Even those that have sophisticated data warehouses and data lakes can struggle to federate their data assets in such a way that they have all their enterprise data in one place -physically (a traditional data warehousing concept), or virtually (via a data mesh) - to derive insights from. The ability to share and integrate enterprise data is a critical component for sustained AI success. Effective communication that traverses the value chain enables seamless sharing and integration of data across departments – or at least, exposes risks, gaps, and misalignments between them. This helps in the creation of a true enterprise dataset for AI consumption. Communication between departments ensures alignment in enterprise data governance practices. This includes the ability for the company to address data security, data privacy and protection, risk management, and compliance in a truly holistic manner.

Getting Started

With an enterprise AI strategy defined, the CEO and the C-suite will need to align and cascade the AI-enabled enterprise strategic objectives with the objectives of individual departments, and cascade them into departmental goals. The executive leadership team will then need to collaborate to coordinate the cascading of the enterprise AI strategy and the derived departmental goals, tailoring their messaging to their individual departments. A commensurate company communication plan - including communicating department goals to each department - should be developed in parallel to the development of the department goals, or not too long after. During this departmental communication, it will be important for leaders to ensure that they refine their messaging to teams and/or individuals who would potentially be directly involved in AI-related efforts. These leaders will need to communicate how AI initiatives support their specific department's goals while contributing to the overall corporate vision and objectives. When developing your departmental communication, it will be crucial for department leaders to tailor communication to anticipate and proactively address specific concerns, risks, opportunities, and benefits of AI implementation for their departments. Leaders should focus on highlighting how AI can address departmental pain points, create operational efficiencies, alleviate rote manual work by automation, or enhance existing processes. Leaders should focus on two aspects of sound departmental communications: i. making the information relevant to an employee in that department, and ii. making potentially esoteric or nebulous AI concepts and their applications more relatable to an employee in that department.

Similarly, the messaging will need to be adjusted for those teams and/or individuals that are not immediately involved with AI-related initiatives within a department, or within entire departments. In this scenario, it will be vital to reiterate that although specific teams/individuals might not be involved in AI-related efforts at that specific time, they too have a role to play to contribute to the success of the enterprise AI strategy. It will be important to ensure that teams/individuals who are not immediately involved with AI-enabled goals do not feel left out, nor should they feel the work belongs somewhere else and that they have no stake in its enterprise-level success.

The CEO and C-suite/leaders of departments should consider a plethora of events as part of their communication activities. Note that contingent on the size of your organization, some, or all these activities might be less important than having meaningful and engaging dialog with your teams/employees. This is not to say that these types of engaging, smaller group/more discussion-focused sessions should not be considered within larger firms. These types of activities, when conducted at the team level, can be more meaningful to employees than broadcast-type communications. Personalization of the messaging – whether discussing AI strategy or any other broad enterprise initiative – will always be more successful than a purely mass communication type of communication strategy. Smaller group forums help leaders synthesize information for employees, and allow employees to ask questions that they might be hesitant to ask in a broader forum. This establishes an invaluable feedback loop.

Broader activities that C-suite leaders should consider encouraging cross-functional collaboration include organizing workshops, or forums. When enterprise AI strategic objectives traverse multiple departments - and all well-crafted enterprise-level objectives should - team-building activities centered around common AI objectives will prove invaluable to dissolve departmental silos. As discussed earlier, an enterprise AI strategy will fail when subjected to corporate departmental silos. These activities will help to facilitate the exchange of ideas, enabling departments to collaborate and work well together on AI projects that span across the value chain. Thematically similar to Data Strategy and Governance concepts, wherein, departments nominate "data stewards" to champion sound data practices within their departments, department leaders should consider designating individuals from each department as AI strategy liaisons. These liaisons act as communication bridges, relaying information about AI initiatives, progress, and opportunities for collaboration between their department and other departments, as well as assist in ensuring ongoing alignment to the broader enterprise AI strategy. Figure 23 below presents a recommended profile of a departmental AI representative.

AI Representative – Recommended Profile:

GOAL: Selecting the appropriate AI representative is crucial for ensuring alignment with the overall enterprise AI strategy. The chosen representative should ideally exhibit leadership qualities, possess solid communication skills, and have a commitment to fostering AI education, governance, and experimentation within their respective department. The individual will champion their department's alignment with the broader enterprise AI strategy as well as their departmental goals as pertain to AI.

a. Current AI Practitioner and/or Early AI Adopter: someone within the department currently using AI.
b. Influence/Leadership: Holds influence or can influence within your department.
c. Communication: Ability to effectively communicate concepts to non-technical staff (with coaching and education).
d. Leadership/Collaboration: Ability to collaborate cross-functionally.
e. AI Knowledge: Basic AI understanding and familiarity with AI tools and technologies relevant to the business domain.
f. Vision for AI Implementation: Ideas or strategies on how AI can positively impact the department and our company.
g. Commitment to AI Governance and Compliance: Willingness to ensure AI initiatives align with policies and standards.
h. Availability and Commitment: Availability to participate in scheduled AI meetings/workshops and commitment to representing your department's interests within the consortia. Expected time commitment is: <FILL IN YOUR ANTICIPATED TIME ALLOCATION>.

Figure 23: Recommended Profile for a departmental AI representative

These AI representatives can potentially be the glue that holds your enterprise AI strategy's execution together across your departments. They are concurrently expected to serve as the AI strategy ambassadors, departmental goal evangelists, AI governance liaisons, ideation and experimentation point of contact, and cross-departmental conduits, amongst a host of other evolving functions. It is important to ensure that the departmental AI representatives are set up for success, such that they can make the department they represent achieve success with enterprise AI. It stands to reason that that organization invests additional time in providing support, resources, and additional education to these AI representatives. Figure 24 below presents additional support that a company can provide to their nominated department AI representatives.

AI Representative – Extra Support Provided:

GOAL: To equip nominated representatives with the (additional) support, knowledge, and skills required to effectively champion AI education and governance within their teams, fostering a culture of responsible AI adoption.

a. Review of AI fundamentals: Review of AI in simple terms without technical jargon, current examples of AI applications relevant to various business units, highlighting advantages and potential challenges of AI adoption in the workplace.

b. Role of AI Representative: Outlining their role in promoting AI education, governance, fostering an environment conducive to experimentation, and helping the department meet stated AI goals. Overview of their participation in the value chain and how they can collaborate with their peers from other departments for enterprise AI initiatives.

c. AI Governance and Ethics: Recap AI Governance policy and exploring the significance of AI governance for ethical and responsible AI use.

d. Compliance and Regulations: Understanding relevant regulations and guidelines for AI implementation in our industry's context.

e: Education feedback: Solicit feedback and augment techniques for creating ongoing engaging and informative AI training sessions tailored to non-technical audiences.

f. Cultivating a Culture of Experimentation: Encouraging innovation in adopting new AI solutions, discussion on embracing a mindset that views failures as learning opportunities in our experimentation.

g. Knowledge Management: Review internal Knowledge Management forums, where they can collaborate, share knowledge, and seek advice.

Figure 24: Recommended additional support for selected departmental AI representatives.

The C-suite should consider establishing a series of workshops. Workshops or collaborative sessions should include the AI liaisons and representatives from different departments. Note that depending on size and scale of your company, the AI liaison and the department representative could be the same individual. Representatives from departments that are required for specific AI initiatives as prioritized by the enterprise AI strategy should be required to participate.

However, regardless of company size, you might want to consider having individuals representing departments that are not immediately involved in AI activities participate as well. This will build a culture of AI inclusivity, and ensure that their voices are also contributory to true enterprise-level initiatives. These sessions should be ideal opportunities for department leaders to communication the enterprise AI strategy, educate employees on AI risks and opportunities across your industry, share and gain insights, collect actionable feedback from employees, and develop potential areas for collaboration between departments.

Once the initial series of communications that introduce the enterprise AI strategy and the department-level objectives that are reflective of this enterprise strategy are complete, companies should move to operationalize regular communications regarding AI progress across the enterprise. These communications will include bespoke messages to the range of stakeholders across the organization, but we will focus on employee (internal) communications here. Operationalization of communications would typically mean that ongoing communication regarding AI would piggyback on regular interdepartmental meetings, featured as an agenda item. However, with broader enterprise AI projects, there will likely be a variety of forums that report on AI progress. These include sprint reviews within the Agile Scrum framework, as well as other Agile ceremonies, stakeholder, and sponsor meetings, etc. It is likely that – at least for the first couple of years of AI strategy execution – stakeholders and key resources across the value chain will convene on a regular manner for interdepartmental meetings focused on AI strategy. These meetings would typically cover program updates, progress, and a recalibration of shared objectives, and would feature departments sharing their AI-related accomplishments, challenges, and potential areas for collaboration, under the auspices of the enterprise AI strategy and departmental goals. Ongoing communications will also expectedly leverage established corporate communication vehicles. This omnichannel approach would encompass leadership emails, stakeholder email updates, newsletters, intranet, internal social media, Knowledge Management platforms, collaboration platforms, etc. Expectedly, these channels will feature AI success stories, case studies, or relevant articles that illustrate the benefits of AI across departments, clearly tying them to the enterprise AI strategy.

Chapter Sixteen: Enterprise Best Practices – Basic Principles: Part 4

After having reviewed the Basic Principle of Communication across the Value Chain in the previous chapter, this chapter will explore the final of the four Basic Principles, Basic Principle 4: Continual Measurement, Continual Improvement, as depicted in Figure 25 below:

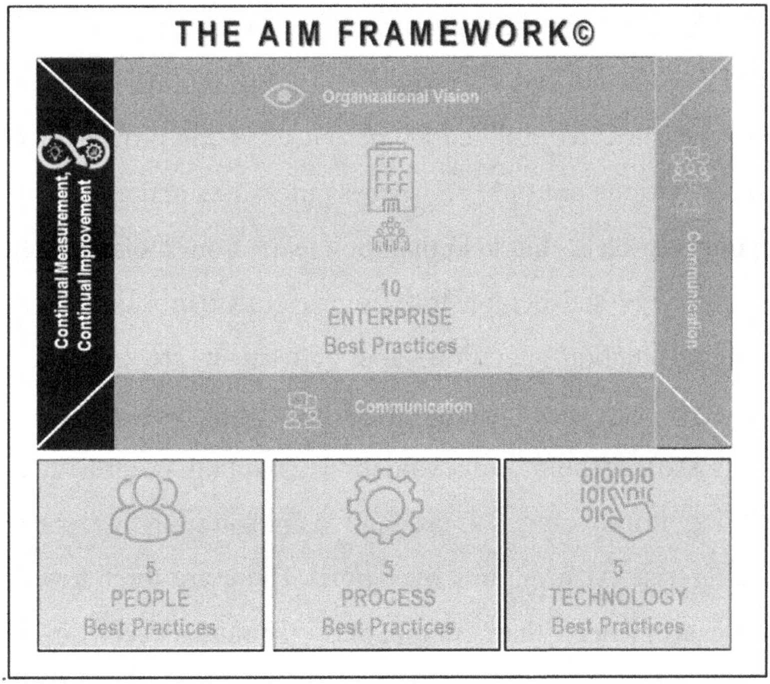

Figure 25: Basic Principle 4: Continual Measurement, Continual Improvement

iv. Basic Principle 4: Continual Measurement, Continual Improvement

The Basic Principle of "Continual Measurement, Continual Improvement" is shown as a vertical bar towards the left of the image as depicted in Figure 25. An established culture around continually measuring the success of AI programs, and investing in continually improving them, is critical for ensuring sustained AI success. Benchmarks serve as critical reference points for continual measurement and continual improvement in enterprise AI initiatives. Stakeholders across the spectrum will expect pertinent success measures to ensure that AI programs are proceeding to achieve the success measures as outlined at the outset of the project. It is vital that organizations not only set success measures and gauge progress on an ongoing basis, but also that they learn from their opportunities and/or missteps and demonstrate improvement in a continual loop. Ongoing evaluation, iteration, and betterment are key factors in assuring the long-term success of AI initiatives. Benchmarks are an essential starting point for evaluation, enabling progress tracking, identifying improvement areas, and demonstrating the success of your enterprise AI programs over time. There are three key aspects of measurements that companies should keep in mind. Measurements should be understandable, reproducible, and intentional.

1. Understandable: Measurements must be understandable to be effective. If people cannot understand the characteristics that are

being measured, the measurement will not help to reduce uncertainty or be useful, despite whether the object being measured is important.

2. Reproducible: Measurements must be reproducible. The main reason for focusing on instruments and conditions of measurements, is to make sure that you are able to produce consistent measurement results, and to understand any factors that might introduce variability into the measurement. If there is no trust in the consistency of the instrument, or an ability to make allowances for variables in conditions, then the measurements themselves cannot be trusted or might have very little meaning.

3. Intentional: Measurements must be intentional. This implies that in order to measure, you must have two objects that are the same, and that meaningful measurement will characterize the differences between them. There needs to be a specific and stated reason for measuring the things that you are measuring.

The expected value created by investing efforts into *Continual Measurement, Continual Improvement*, as well as how a company might get started, are outlined below:

Continual Measurement, Continual Improvement - Expected Value

Expected Value 1: Baseline, Then Make It Better

Establishing some foundational baselines is vital for continual measurement and continual improvement in the successful execution of enterprise AI programs. Baseline benchmarks are an important point of reference for any company that has operationalized, or is seeking to operationalize, their AI strategy. Baselines allow companies to be able to showcase their enterprise AI progress and highlight their successes in their AI initiatives. Demonstrating improvements over time against these benchmarks helps in illustrating the impact and value of AI programs to stakeholders and decision-makers. As AI technologies evolve rapidly, continual measurement and continual improvement are essential to keep pace with AI advancements.

For the CEO and C-suite, baseline benchmarks provide valuable insights for decision-making regarding what aspects of their enterprise AI strategy they must prioritize and/or reprioritize. Benchmarks help drive resource allocation, prioritization of investments, or even adjustments to the overarching AI strategy itself. For companies that claim to be, or aim to be, data-driven enterprises, baselines are a crucial tool in helping render informed decisions based on performance data, stated tangible metrics, and measurable goals. Continual measurement can help in the identification of your process inefficiencies as AI-enabled enterprise business objectives are operationalized. This identification helps in facilitating process optimization through AI, leading to enhanced operational efficiency across your company's value chain. Baseline benchmarks serve as a basis for iterative cycles of improvement, allowing your cross-functional teams to ideate, test, and implement enhancements while comparing against established performance baselines. The iterative nature of AI development demands ongoing evaluation to adapt to changing business needs and technological innovations. Continual improvement in your enterprise AI program will only be facilitated by virtue of iterative refinement.

Expected Value 2: Technical Table Stakes

I like to think that AI models, like expensive cars that depreciate as soon as you drive off the automobile dealer's lot, depreciate as soon as they are operationalized. Just like the fact that some car models are simply expensive cars, and others are considered classics or vintage cars, AI models require care and maintenance soon after they are implemented. AI models can be notoriously unpredictable when they are "in the wild," establishing unexpected correlations. Models tend to "drift" and "hallucinate," both concepts that shall be examined as we delve into Explainable AI (XAI).

Models can also suffer degraded performance over time and/or when larger data sets are introduced, with outputs that took seconds during training - or soon after implementation - taking endless hours as real data is introduced, or the size of the data set is exponentially larger than what the model was trained with. Model maintenance is therefore considered to be AI technology table stakes. Continual measurement will allow your firm to assess the effectiveness of AI applications that have been prioritized as part of the enterprise AI strategy, against predefined performance indicators. A continual measurement mindset will help in surfacing AI project challenges, limitations, and opportunities, and will therefore assist in helping make focused improvements to your enterprise AI program. Ongoing iteration and refinement of AI models based on insights gained via continual measurement leads to improved accuracy, reliability, and predictive capabilities. In turn, establishing feedback loops between model performance and improvement drives continual enhancement of AI algorithms.

Continual measurement ensures ongoing assessment of data quality, relevancy, and suitability for AI models. This ongoing assessment is vital in preventing data drift and/or degradation. Continual measurement that occurs in real-time, or near real-time can help in flagging shifts in data patterns. This can help your firm to make immediate AI model adjustments to maintain the accuracy of your AI models. Baseline benchmarks therefore serve as a reference point to evaluate the initial performance of AI models or systems. They provide a starting point to measure the effectiveness, accuracy, and efficiency of AI models. Benchmarks enable comparison between different iterations or versions of AI models. By comparing against established benchmarks, it becomes easier to track progress, identify improvements, or detect potential regressions in the performance of AI models over time.

Expected Value 3: Moving the Cultural Needle

Continual Measurement and continual improvement, when intentionally implemented as part of your enterprise AI program, will greatly help in fostering an atmosphere of ideation, innovation, and a culture of experimentation throughout your organization. It sends a clear and unambiguous message to your employees that the focus – after setbacks or failures – is intended to learn, grow, and get better, and not on punitive measures or castigating employees for these setbacks or failures. This can help encourage individuals across your value chain to experiment and innovate, seeking new or better ways to augment AI into the firm as driven by the enterprise AI strategy. It promotes a learning culture, creating a safe space for teams to experiment, learn, and innovate without fear of failure.

Enterprise agility will be crucial in the Age of AI. Agility – whether in the context of Agile Scrum, or Scaled Agile (SAFe), or enterprise agility (a natural extension of agile tenets), can serve as the linchpin for companies that will face significant AI-fueled changes in the next few decades. These AI-fueled changes are likely to disrupt and transform entire industries, potentially rendering them unrecognizable from what we know them as today. In this era of dramatic and rapid changes, enterprise agility will ensure that an organization can survive and thrive amidst this changing environment. A continual improvement approach will help your company build its enterprise agility muscle by allowing rapid company shifts to meet changing market conditions, and leverage the latest advancements in AI. This will help your company be able to adeptly manage new risks and capitalize on new opportunities. The iterative nature of enterprise agility will help the firm become much more adaptive in nature. This flexibility and adaptability will be invaluable during a rapidly evolving AI landscape. Enterprise agility is less about a process or a tool, and more about an organization's culture. Continual measurement and improvement will help in moving the corporate cultural needle.

Expected Value 4: Quality Assurance and Quality Control

Although the concepts are often misused interchangeably, quality assurance and quality control are separate and distinct facets of AI development, both being equally critical in ensuring a high-quality product. Quality control is a multi-step review process that is conducted by the AI project team, and focuses on the identification of defects throughout the lifecycle of a project. It is said to be a reactive process to verify the quality of a product. Baseline benchmarks highlight areas where AI systems may fall short or underperform compared to the desired targets. They help in pinpointing specific weaknesses or inefficiencies, guiding efforts for targeted improvements or optimization strategies.

Quality assurance on the other hand, is a broader, process-centric, proactive process that occurs throughout the AI development process. Quality assurance helps in the management of quality, and focuses on the prevention of defects. Quality assurance reviews the entire system holistically, and occurs after quality control (which tests individual parts of the system), but before operationalization of the AI system/model being built. Baseline benchmarks play a crucial role in quality assurance and validation processes for AI systems by helping to validate whether the implemented changes or updates have led to desired improvements, or maintained the system's target performance standards.

Expected Value 5: Regulatory and Compliance

As established early on in this book, ensuring that AI models are free of inadvertent bias and proxy discrimination is a top concern, and one of the key factors behind Explainable AI (XAI). This concern is agnostic to industry or company, and regardless of the specific type of AI in the sprawling field that is AI. Regulatory and Compliance shall be examined further in this book, but it is important to understand that the vibrancy and fluidity of the overall field of AI has disallowed regulation to be developed. Therefore, it is challenging for organizational compliance policies commensurate with these absent regulatory guidelines to be developed. It is even more challenging to build corporate policies that can stay on par with the rapid evolution of AI. While the European Union issued guidelines in 2023, and President Joe Biden issued an executive order on AI, there are few industry-specific regulations governing entire sectors. Industries, by-and-large, have still been left to decipher the implications of evolving regulatory frameworks, and their compliance to these frameworks.

That said, most companies that can be considered as the "early adopters" or the "pragmatists," have sought to establish their AI practices around sound ethical principles. This includes ensuring that AI models are free of bias. Continual measurement and continual improvement are a key part of bias detection - and subsequent mitigation - in AI models, therefore ensuring fairness of AI models and ethical deployments of AI systems. The insights that are gained through continual measurement drive the formulation of robust ethical frameworks, fostering responsible AI practices. As the regulatory landscape continues to evolve, a continual measurement practice also ensures that your organization can swiftly align with new/updated regulations and thereby mitigate risks of non-compliance.

Expected Value 6: Being Data-Driven to Improve Being Data-Driven

Claiming to being a data-driven company has entrenched itself in the corporate lexicon. Regardless of sector or industry, most companies are built on the foundation of data, the only distinction being the volume and quality of data to enable enterprise decision-making. Many organizations claim that they are data-driven and seek to make data-driven decisions. This "data-driven" mindset, however, has not been ingrained across an enterprise. The siloed approach towards digitization, and the resultant data from this digitization, has yielded mixed results, wherein certain departments within a company can stake their claim to be more data-driven than others. For instance, in your firm, with a robust Customer Relationship Management (CRM) infrastructure – whether enabled by Salesforce or Microsoft Dynamics – your sales team could claim that they are a data-driven department. The same might not apply for product areas that maintain their own unique customer database. The records of this unique customer database might not reconcile with the data residing in your CRM system. If you consider your CRM system to be the golden source for your customer records, you can see why this might pose a challenge. The truth of the matter is that most organizations, when they claim that they are data-driven, are likely referring to a part of their firms, and not necessarily their entire enterprise.

This is where companies would do well to invest upfront in defining what it means to them for truly being "data-driven." Companies should invest upfront time in identifying what their strategies are, the desired outputs, and the outcomes that they are seeking from these strategies. To continue securing funding, being able to demonstrate value to internal and external stakeholders - such as shareholders, the Board of Directors, and employees - the C-suite has to be intentional in measurement of the success of their enterprise AI programs. There are a whole host of measures available to choose from – from measuring operational efficiencies, cost savings, customer satisfaction, etc. All these measures are critical and are completely driven by an organization's vision - and specifically how this enterprise vision manifests itself into their enterprise AI program.

It is recommended that firms pick five to ten most important Key Performance Indicators (KPIs) to focus on, establish baseline benchmarks at the outset (or use existing measures as benchmarks for ongoing programs), and incorporate achievement of these KPIs across employee objectives, including those of the C-suite performance goals. To ensure that the enterprise continues to strive towards achieving the broader vision for enterprise AI, it will be important to ensure transparency and accountability by demonstrating the efficacy and effectiveness of the progress and highlight progress. A maturity model for AI is presented as one of the ten enterprise-level best practice recommendations in this book. This maturity model is part of what is described as the AI Body of Practice, the "AI BoP." The "AI BoP" is one facet of this continual improvement mindset. While the "AI BoP" prescribes a "Center of Excellence" (CoE) type framework to measure the maturity of an organization's AI practice, establishment of KPIs and benchmarks evaluates and measures how this practice is adding value at the enterprise-level.

Chapter Seventeen: Ten Enterprise-Level Best Practices – Part 1

Having examined the four Basic Principles of the AIM Framework©, this chapter will commence an exploration into the ten Enterprise-Level Best Practices, as depicted in Figure 26 below:

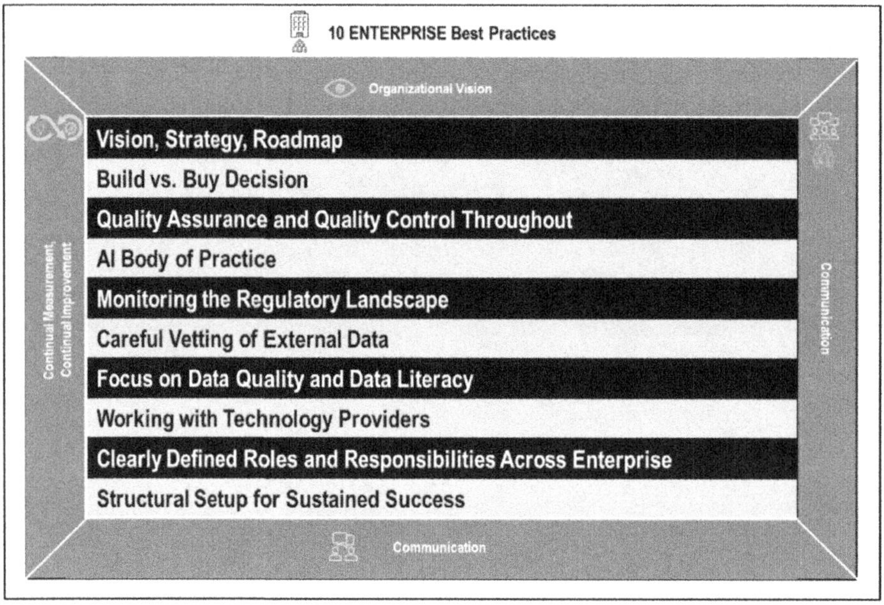

Figure 26: Ten Enterprise-Level Best Practices

1. Enterprise Best Practice 1: Develop an AI Vision, Strategy, and Roadmap at the Departmental Level {⚡ KEY RECOMMENDATION}

Once enterprise objectives are developed/augmented to incorporate AI into your corporate goals, organizations should invest time and attention into developing a vision, long-term strategy, and roadmap - as pertains to AI - for each department/division within your company. Within our context, departments are also synonymously referred to as divisions, business units, or functional areas. These include IT, Marketing, Financial, Product Development, Manufacturing, Sales, etc. Most organizations across industries generally have the same/similar functional areas, with some departments unique to their particular industry. Across industries, their functional areas might carry out the same functions, but might be called something else. Departments will all have quite diverse use cases within their functional areas. There will exist opportunities for significant overlap in terms of these use cases across departments. Starting with each department and then socializing each departmental AI strategy to discern commonalities, is an ideal bottom-up complement to the top-down enterprise AI strategy. Note that there are distinct differences between the enterprise AI strategy and commensurate departmental goals that support this corporate AI strategy, and a department-focused approach for AI. While not all departments might initially participate in prioritized projects to support the enterprise AI strategy, each department within every firm has significant opportunities to capitalize on opportunities with AI.

Building a 360° AI Strategy

There are a multitude of reasons that your enterprise AI strategy must be complemented with a department-based strategy and roadmap for that specific department. This department strategy should be inclusive of whatever aspects of support for the enterprise AI strategy that the department is contributing to. In other words, departments across your enterprise value chain will have objectives that have been cascaded down from the enterprise AI strategy.

First, note that not all departments might have these cascaded objectives – depending on what use cases have been prioritized at an enterprise level. Secondly, it is important to note that those departments across your enterprise value chain involved in supporting prioritized enterprise AI use cases, will have varying degrees of participation and contribution to the enterprise AI strategy. The extent and depth of a department's contribution to the enterprise AI strategy will change over time. This is depicted in Figure 27. The enterprise AI strategy sits horizontally across the top of the figure, covering the entirety of your organizational value chain. Imagine if your organization has eight departments. These departments/functional areas/divisions are depicted as vertical white boxes across the value chain, and include (from left to right) Product

Development, Marketing, Sales, Finance, IT, Customer Service, Human Resources and Legal. Consider that your organization has decided to prioritize one AI use case in support of the enterprise AI strategy. The impact of this enterprise AI strategy, at a divisional level, will require all departments except Finance, Human Resources, and Legal. The departments that will require to contribute to the prioritized AI use case as defined in your enterprise AI strategy, and depicted as vertical white bars – Product Development, Marketing, Sales, IT, and Customer Service - each have a vertical black bar that inherits from the enterprise AI strategy horizontal bar. It would be highly unlikely that every department contributing to this enterprise strategy will be required to invest the same amount of effort. Therefore, some black bars are deeper into some departments such as IT and Product Development than others. Which of these departments are required to participate towards the enterprise AI strategy, and at what extent, can and will likely change from use case to use case, and might look different on a periodic basis as the enterprise AI strategy is refreshed commensurate to shifting business objectives. It is important to note that although business objectives and associated AI enterprise objectives might periodically shift, it is unlikely that a company's broader business vision and long-term strategy, and hence, the broader AI vision and long-term strategy, would dramatically change in a given time period.

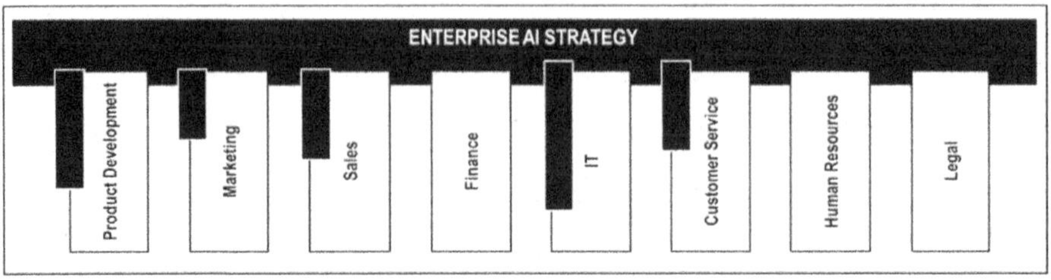

Figure 27: How the enterprise AI strategy might manifest across the value chain

Third, there will be use cases that the CEO and C-suite have prioritized that will either be completed, be deferred, or reprioritized on a year-to-year basis. As the organization shifts in accordance with changing business conditions and advancements in AI, there will be new use cases that are vetted and reprioritized or deprioritized. There might even be some active AI projects that get put on the backburner or shelved entirely. The support of each department towards this enterprise AI strategy will consequently be a direct reflection of what active use cases are being worked upon, and those that are on the immediate horizon. The prioritization of enterprise AI uses cases should be directionally in support of the enterprise vision and objectives around how AI enables the long-term business vision and business strategy. As business strategies evolve, so should the enterprise AI strategy, and as such, the need for departments across the value chain to support the execution of the enterprise AI strategy.

As has been stated, the Age of AI is poised to impact, influence, disrupt, and disintermediate every facet of a company's value chain. While it is highly unlikely that the prioritized list of use cases from your enterprise AI strategy will concurrently envelope every single facet of your value chain, AI's impact on the business of a specific business unit will certainly be felt. Focusing on the top prioritized AI use cases as outlined in your enterprise AI strategy is appropriate. Undertaking too many concurrent enterprise-level AI use cases will risk diluting the focus of the enterprise and risk stretching your organization too thin. However, by solely focusing on the prioritized set of enterprise AI use cases, you do not want to create AI disparities within your firm. Since not all business units will be concurrently involved in enterprise AI initiatives, you do not inadvertently create an organization of "*AI haves and have nots*." This would be a derivation of the misstep that organizations across industries have made in effectuating their Digital Transformations. You do not want to foster an environment where, if you only focus on prioritized AI use cases and AI-enable a portion of the business units contributing into these use cases, you implicitly risk other business units being left behind.

A Holistic Approach

Additionally, it should be noted that solely an enterprise-level set of use cases that involves the active participation of a particular business unit will not AI-enable the business function of that department. Any department involved in the enterprise AI strategy will likely only see a portion of how the division operations change and become AI-enabled. These AI-enabled changes are also likely to transform the core of the business unit's functions – they will transform how that business unit interacts with other business units across the value chain. A business unit AI vision and strategy is not the same as an enterprise-level strategy. Your firm will need both - a top-down enterprise AI strategy that cuts across the value chain, and a bottom-up department-level AI vision and set of priorities around how the business unit can transform by capitalizing on AI. A department-level AI strategy does not supplant your enterprise-level AI strategy, it supplements and complements the enterprise strategy.

Figure 28 visually depicts what a business unit AI strategy might look like. In this instance, we use the same example of the prioritized enterprise AI use case that we selected in Figure 27. The notable differences here are that each business unit has an AI strategy commensurate to the business function of that department, depicted here as a translucent cylinder that encases the department, as well the department-level objectives that are cascaded down from the enterprise-level AI strategy. Figure 28 illustrates what a holistic AI strategy looks like for your organization. While there might be "n" number of enterprise AI strategy use cases for your company, along with their associated department-level objectives that cascade from top-down, each department will have one and only one departmental AI strategy with a specific set of objectives around it. Every departmental AI strategy must be directionally aligned to the enterprise AI strategy. The departmental strategy should be congruent with where you envision that line of business is headed in the Age of AI. For instance, using the example depicted in Figure 28, the Human Resources (HR) department is not involved in the prioritized enterprise AI use case.

However, HR will significantly transform and evolve in the next several decades. Perhaps, one of HR's business objectives is to streamline the applicant's experience. A departmental strategy for HR should include how AI can be deployed to automate facets across the applicant recruitment value stream. This could not only add value to applicants, but also allow for operational efficiencies, reduce friction, and result in cost savings. The HR strategy could potentially include AI-enabled screening and filtering resumes to human recruiters. This has been a real use case with very real ethical considerations and potential for issues as we will explore later in this book. One of the key considerations of any HR strategy should be to try and understand how your business will change over the next decade, and what your enterprise talent profile looks like in the next ten years. HR will need to understand implications to jobs and advise the C-suite appropriately. The C-suite and HR will need to commence developing plans to skilling/reskilling/retraining employees to be ready for the jobs of the future. This should feature as a part of any HR AI-enabled business strategy. The point here is that even for a division not immediately and directly involved in supporting the enterprise AI strategy, HR itself will need an AI-enabled business strategy that works seamlessly to further the enterprise AI vision.

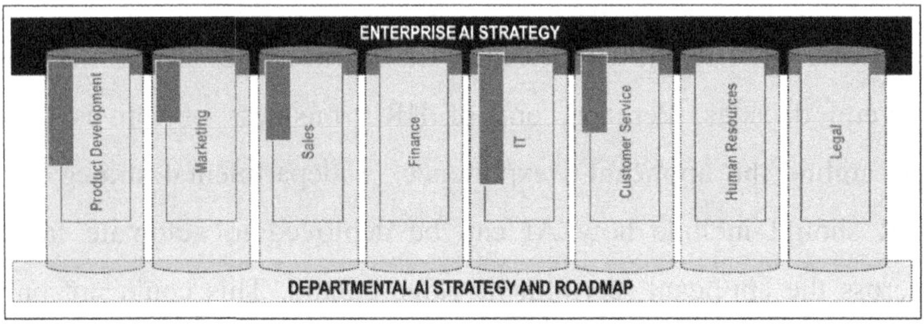

Figure 28: Departmental AI Strategy

Developing a Holistic AI Strategy

With AI expanding at a rapid pace across industries, the resulting "bandwagon effect" has disallowed firms from catching their breath, taking a step back, and formulating and articulating a visionary long-term plan for how AI can impact their enterprise, much less their functional areas. Even with the "early adopters" and "pragmatists," there are likely only a few firms with a clear line of sight into the current state of how AI is going to impact any one given functional area within their company in relation to the envisioned future state of where they'd like to go with AI within that function. With organizations that have been late to the establishment of their AI practice, having to play catch up has proven to exert pressures against being able to take a long-term strategic look at their plans, ceding that planning for expeditious implementations. This visioning exercise should be a cross-functional effort between representatives across the company. This exercise can also be conducted by technology consultants in partnership with a firm. Technology consulting firms can be very helpful in that they can bring in timely and topical industry knowledge to help shape and craft a firm's functional area strategic roadmap.

Regardless of who helps in crafting and executing on this strategy, the C-suite should start with the simple question – "what business value are we looking to derive from this specific functional area?" This business value should be clearly in furtherance of corporate goals and objectives and highlight three to five key desired outcomes, such as "improving the customer's experience," "faster time-to-market," "decreasing operating costs," "gain a competitive edge by virtue of x," etc. Ideally, companies should have at least a five-year roadmap for departmental AI programs, and work backwards from that envisioned point in the future. A potential forward look into the future and evolution of your departmental AI strategy does not need to be grandiose. Depending on your company and objectives, this could be everything from expanding on what a functional area offers today to something aspirational.

The formulation of a departmental vision and strategy for an AI-enabled business unit should be led by the department head/line of business head. Typically, this individual will also have a seat on the executive leadership team of the company. The business unit head, as a member of the C-suite, would be the appropriate lead to develop the business unit AI strategy for several reasons. First, the business unit head would have had a role in crafting the enterprise AI strategy. Implicitly, they should clearly understand the enterprise AI strategy, and the direct correlations to their business unit. The C-suite, line of business head would also have a clear line of sight into the role of the business unit in supporting the enterprise AI strategy, and the departmental objectives that cascade down from the enterprise AI strategy that they helped shape. On a more fundamental level, a C-suite business unit head should have a clear understanding of how their business is transforming due to AI. They need to be able to champion an evolution of their own business, setting a vision and strategy for their unit that they can then articulate to their peers and their teams. A fundamental aspect of the C-suite's responsibilities is to get their teams and stakeholders excited to share in the execution of this strategy. This Enterprise Best Practice recommends that the C-suite line of business heads lead the development of their departmental vision and strategy, or refresh the same, predicated on AI.

The C-suite will likely benefit from involving a group of individuals in helping them build this departmental vision and strategy, including their direct reports, and potentially outside consulting firms. Consulting firms can add significant value by providing C-suite leaders with a company-agnostic industry-level perspective on AI's transformative potential for their specific business unit. As the strategy is being developed, and then pivots from formulation to execution, the C-suite leader would benefit by relying on their designated AI representative for their department for being the point person to ensure alignment between the departmental strategy and the enterprise AI strategy. Not only can the AI representatives help with being able to connect the dots between the overall enterprise initiatives and their particular department, but they can also help bridge the gap between "shadow AI operations" and the enterprise AI program. A well-crafted departmental AI strategy will more than likely encompass AI work that originated within the department before the departmental AI strategy was crafted. This will ensure that AI activities within a department are not "one offs," and can trace their activities to overarching departmental and/or enterprise AI objectives. The departmental strategy and the enterprise AI strategy must work together in harmony, and it is the responsibility of the divisional AI stewards/representatives, project teams, and divisional leadership to ensure this synchronicity. Ultimately, it is the C-suite/line of business head for each department who is accountable for ensuring that this balance is maintained. Enterprise Best Practice 1 espouses the need for a departmental vision and strategy around AI. In combination with the Basic Principle 1 that speaks to an enterprise AI vision and strategy, this best practice is foundational to the AIM Framework©, and implicitly, to the success of your AI program.

Chapter Eighteen: Ten Enterprise-Level Best Practices – Part 2

The second Enterprise Best Practice, "Build vs. Buy Decision," tackles the fundamental choice that all companies grapple with when building out - or refreshing - their enterprise AI programs. This choice is known as the "Build vs. Buy" decision - whether to ***build*** out your own AI models in-house and using your own data for these models along with data sets purchased from trusted external providers, or to ***buy*** services from technology providers/third-party vendors, wherein the vendors build, manage, and maintain the AI models, as well as manage the data that fuels these AI models. Note that any given company could have "n" number of AI models as part of their enterprise AI strategy. These models in turn could be fed by multiple data sources, including internal data, publicly available data sets, as well as data sets purchased from third-party data vendors. This "Build vs. Buy" decision holds significant implications for your organization's AI strategy, scalability, resource allocation, and the overall success of your AI program. Figure 29 provides an overview of some of the primary factors that companies consider when deciding between a "build" approach versus a "buy" approach.

	BUILD	BUY
Strategic Alignment	• Allows closer integration with company vision and strategy. • Increased alignment with evolving business objectives. • Nurtures enterprise innovation and builds AI expertise. • Requires continual AI investment.	• Allows closer integration with industry direction as pertains to AI. • Better alignment with short-term enterprise objectives. • Might create misalignment between enterprise goals and vendor goals. • Creates a dependency with external third-party providers.
Customization	• Great level of control and customization to your enterprise vision. • AI models can be bespoke to your specific organizational needs. • Considerable upfront and ongoing AI investment. • Not a "quick fix" – not ideal for faster time-to-market.	• Faster time-to-market, potentially turnkey solutions. • Inability to significantly customize solutions to your enterprise vision. • Ability to leverage expertise and technology without need for expertise. • Ability to keep up with AI advancements and new data sources.
Resources	• Requires building out internal expertise and domain knowledge. • Requires investment in talent acquisition and retention for AI. • Needs investment in skilling/reskilling teams. • You might be challenged to find/retain skilled talent.	• Tap into external expertise, keeping up with AI evolution. • Leverage the skills and knowledge of specialized AI vendors. • Lesser need for investments in hiring and training. • Reliance on vendors might pose challenges for control and alignment.
Scalability	• Allows to develop solutions that align precisely with evolving needs. • Flexibility in scaling AI infrastructure and models to changing needs. • Might have challenges of resource allocation and expertise availability. • When to scale and how much to scale is within your control.	• Provides immediate scalability as vendors offer scalable solutions. • Ability to quickly adapt to increased or fluctuating demands. • Ability to mitigate extensive development costs when scaling up. • You will be constrained by the vendor's offerings and limitations.
Data	• Greater control over data management and privacy. • You can ensure full compliance with regulatory standards. • Data sovereignty and sensitive information are within your control. • Requires robust data governance and information security measures.	• Involves third-party data management, increased exposure. • Might require access to proprietary data. • Data privacy, security, and ownership will need to be considered. • Contractual agreements and compliance frameworks will be needed.

Figure 29: "Build vs. Buy" decision - some considerations.

There are organizations that fall into the "early adopter" and "pragmatists" categories that typically have an established "build" model, but some lines of business within the enterprise might have gravitated towards a "buy" model. This hybrid model might seem superficially attractive to derive the benefits of both worlds. However, this hybrid model, where some parts of a company choose a "build" approach, and the other chooses a "buy" approach can split your organizational AI focus and will stymie your company from deriving the full benefits of a truly enterprise AI strategy. A hybrid model might be a transient state, where a company is in the midst of transitioning from a "build" model to a "buy" approach, or vice-versa. Most companies have a strong inclination towards one or the other, these preferences having changed over the past few years as technology providers and AI vendors are getting much more robust than they had been in the past.

Enterprise Best Practice 2: Build vs. Buy Decision

{✓ KEY FINDING/RECOMMENDATION}

It is important to reorient on what "build" versus "buy" means in the context of your enterprise AI models, since this definition is slightly different than it would be if applied purely to a software or technology decision. At the most basic, the distinction between "build" and "buy" is who owns the AI/ML predictive models – the companies themselves or their technology providers. These predictive models are at the heart of all AI, being able to make some meaningful predictions with provided data, whether that data is internal, publicly available, and purchased data sets. Organizations have a choice whether to build their own AI and ML models, while consuming external data from data/technology providers. There are no overarching patterns of companies choosing to "build" versus to "buy," but by-and-large, the "laggards" and companies with insufficient internal infrastructure and/or talent, have gravitated towards the "buy" model due to the rapid time-to-market.

Companies with more mature data strategy and governance and data management practices have gravitated towards the "build" model due to their mindset around having mastery of their data, their AI/ML models, and having had the skills internally to support such a build. There are organizations that can be considered as "early adopters" or "pragmatists" that proved out their own AI/ML models internally in a "build" model before deciding to trade that for a "buy" model, partnering with technology providers to achieve scale faster than they could by themselves. These are organizations that have realized that an AI model requires immediate care, feeding, and maintenance, immediately after it has been put into practice. These firms have made the choice to outsource their AI/ML models and focus on their core business, rather than continue investing in building out their AI infrastructure. These companies have also looked at the AI maturity curve and have made the calculated argument that it would be an order of magnitude easier for technology vendors to continue augmenting these AI models with new and improved sources of data. These new data channels will continue to strengthen the AI models and their outputs. Each time a new data element is introduced to a model, the model might require a battery of smoke and/or regression tests, something that some companies are loath to invest time and resources in.

In the "build" model, a company owns total responsibility for building, augmenting, supporting, and maintaining the AI/ML model. In this instance, an organization commits to having the technology, operational infrastructure, and skills in-house to be able to continually maintain and adjust their own models. AI models can tend to drift unless they are beneficiaries to constant upkeep and maintenance. Model drift, also known as model decay, is defined by IBM as "the degradation of model performance due to changes in data and relationships between input and output variables. It is relatively common for model drift to impact an organization negatively over time or sometimes suddenly. To effectively detect and mitigate drift, organizations can monitor and manage model performance as part of data and AI platform." (IBM Watson Studio, 2022).

Even with a "build" approach, companies typically have to partner with data/technology providers in order to procure data to make their models meaningful and deliver any semblance of predictive value. In the "buy" model, a company partners with established technology providers. The companies supply their unique business rules, objectives, and other pertinent data to their provider partner. The vendor/partner/technology provider (terms used synonymously) builds and maintains their proprietary AI models and output/prediction/decision engines. The vendor/partner/technology provider is also responsible for all external data sources that the vendor's proprietary model ingests. The vendor's model ingests these data sources and information that the company furnishes, and produces an output of some kind – a recommendation, a prediction. or a decision, back to the organization pertinent to the AI use case.

There are several such mature and established technology providers across multiple industries. In addition to serving as "full service" technology providers, they also typically serve as sources of external data sets to companies who choose to pursue the "build" approach. Referenced here for illustrative purposes only, is one such technology provider to the Insurance industry, Verisk, and their platform/s "EHR Triage Engine" and "EHR Triage Engine Plus", used within the life insurance underwriting process. Life insurance companies leverage AI and ML in the process of underwriting life insurance policies. These companies can choose to build their own AI models for underwriting (known as automated and accelerated underwriting), or "buy" the service from a third-party technology provider. In the case of the company Verisk, Verisk serves as a prominent provider of external data to many life insurance organizations.

These life insurance companies ingest data from Verisk, and several other external third parties, for their AI models. The AI models then produce a life insurance application decision and/or make recommendations to a human underwriter, who can adjudicate on the application – whether the applicant for insurance will be approved or denied, and what the premium amount should be (what the applicant will need to pay). In this example, the insurance company builds, maintains, and manages their own AI models, while using Verisk as a data provider. Verisk, as explained with the illustrative example, also offers companies the ability to use Verisk's proprietary AI underwriting model. In this case, the life insurance company would provide Verisk with their bespoke, unique, underwriting rules, information about the applicant for life insurance, and other requisite data. Verisk would ingest this data from the company, integrating this data with the multiple data sources that they leverage, and produce an output back to the insurance company. This output could be a recommendation, a decision, etc., based on the contract between the company and Verisk. Verisk offers this service to many life insurance companies. This is the "buy" approach in action. Most life insurance companies leverage Verisk's services across their value chain, and moving from a "build" to a "buy" model, or, launching their enterprise AI program (or at least one facet of it), with Verisk as their trusted partner, would be a logical step for these life insurance companies. Verisk is one of dozens of technology providers offering similar sophisticated services. From serving as data providers to serving as "one stop shops," technology providers mitigate the upfront investment required to enter the AI space. As depicted in Figure 30, Verisk's prospectus states that their platforms can lucidly plug into any life insurance company's underwriting workflow (Verisk, 2020). In addition to mitigating the need for a life insurance company to build and develop their own AI models, Verisk's prospectus for their platform states that engaging with them as a technology provider within the underwriting process offers benefits such as "a dedicated team of biostatisticians and data scientists with expertise in survival analysis and NLP, a global medical research team working with the latest clinical literature and cohort studies, a team of regulatory and data privacy specialists focused on compliance aspects, powerful economies of scale". (Verisk, 2020)

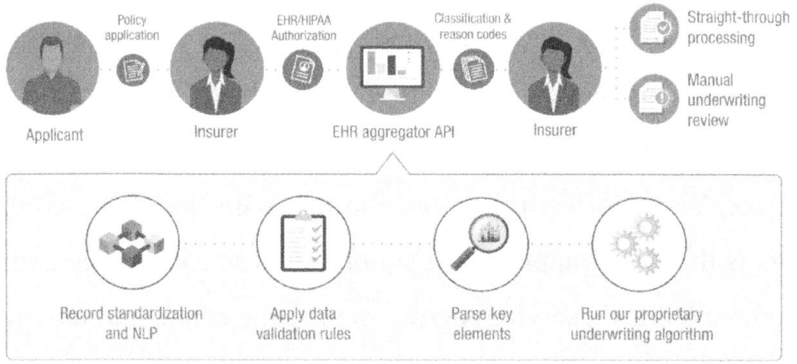

Figure 30: Technology Vendor Verisk and their Automated/Accelerated Underwriting Workflow (Citation in Body).

When organizations decide upon pursuing a "build" versus "buy" strategy, they do so after internal dialog, guidance from their ecosystem partners, and/or recommendations from technology consultants. Some companies decide to extend their ongoing relationship with their technology providers - from being one where the provider supplies external data sources to a carrier's homegrown AI models ("build"), to engaging with the technology provider as a "one stop shop." These companies are typically the ones who have existing, limited-scope engagements with their technology providers, with these engagements being purely of data acquisition in nature.

Once their business models start realizing some success, and it comes time to scale out their AI program, these firms make a conscious decision to expand their relationship with their technology provider. Some firms tend to use the same “build” versus “buy” playbook employed for software or technology decision-making in general, but find that their internal frameworks usually fail to account for the requisite integrations required between an organization and the vendor, regardless of if they pursued a “build” or a “buy” model.

Tools to Facilitate the “Build” versus “Buy” Decision Process

The "Build vs. Buy" decision in AI implementations is multifaceted in nature. The “build” model offers customization and control, but demands considerable investments in building out the requisite infrastructure, securing talent, and resources. The “buy” model, where you can procure AI services from third-party vendors, provides speed, time-to-market, and industry-level expertise. However, this might require your company to relinquish some control and long-term alignment with your unique strategic vision and business objectives. Companies must carefully weigh these considerations, aligning their choice with their strategic vision, scalability needs, data privacy concerns, and resource capabilities to make an informed decision that best suits their AI objectives. There is no “right” approach, but there might be an approach that is better suited for your firm, and your strategic vision, than the other. This decision requires a thorough evaluation of trade-offs. Enterprise Best Practice 2, "Build vs. Buy," provides tools for organizations to be able to facilitate and inform the decision-making process.

The AIM Framework© presents two instruments to assist your company with your decision-making process. These tools, a ***qualitative*** decision-matrix, and a ***quantitative*** decision-scorecard, are aides that work best when used in combination with each other. Note that these aides are just that - aides - and as such they allow for deep customization to your individual organization's unique needs. In fact, the recommendations from these tools would work best when they are tailored to your company's specific objectives. Both these tools, qualitative and quantitative, are intentionally simplistic and generic in nature, so as to allow customization and extension to fit your company's unique business requirements and strategic objectives. These tools are meant as guideposts, and whether they are used individually or collectively, these tools should not be the only way upon which a company bases their "build" versus "buy" decision upon.

A. Tools for the "Build" versus "Buy" Decision Process – The Qualitative Decision Framework

As depicted in Figure 31 below, the "Qualitative Build versus Buy Decision Framework" is comprised of three main sequential decision-making "tiers." The reason for calling these tiers as opposed to steps is because each tier of the decision-making process, unlike steps, can run synchronously with each other, and pursued by different individuals within your company. However, a qualitative decision cannot be reached without all three tiers coming together at the end. Rendering a decision based only on one or two of these tiers would be incomplete.

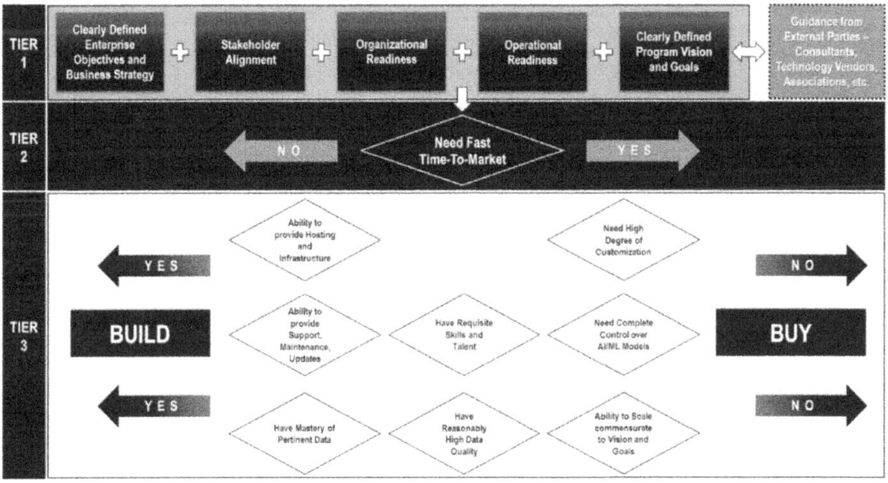

Figure 31: The Qualitative Build vs. Buy Decision Framework

TIER 1

Tier 1 of the qualitative framework consists of five main keys and is not deterministic to either a "build" versus "buy" decision. This tier ensures that the enterprise is ready, willing, and able to commit or recommit to a journey towards being an AI-enabled business. Depicted as the black shaded boxes within Tier 1 in Figure 31 above, all of these five keys need to be in place to guarantee organizational commitment to the holistic enterprise AI program. These five keys are:

1. Clearly defined enterprise objectives and business strategy as relates to the enterprise AI program.
2. Alignment across the internal stakeholder community.
3. Organizational readiness to commit or recommit to the program (People/Process/Technology).
4. Operational readiness to commit or recommit to the program (can the program be supported by the enterprise over the long-term).
5. Clearly defined vision, strategy, and a roadmap for AI within the organization.

This tier is also influenced and informed by external third parties, including consultants, technology vendors, industry trade associations, etc.

TIER 2

The simple "YES" or "NO" decision points represented in Tier 2 are intended to force a decision on what is most important to the organization. For a lot of organizations that are looking to launch their AI programs under external competitive pressures, time-to-market needs to rank very highly in the decision-making process between building and buying. This decision point is therefore represented as the sole decision point for this tier. Your organization is free to replace this key decision with whatever is most important to your company, but it should be noted that time-to-market is a critical consideration when evaluating "build" versus "buy." The AIM Framework©, therefore, strongly recommends that it be featured as a prominent inflection point.

If time-to-market is important and vital for your organization to achieve, then the qualitative decision framework tilts in favor of a "buy" model decision. If time-to-market is unimportant – since a company might already have a relatively mature "build" model already established – then the framework tips towards the "build" model. In either scenario, the decision points enumerated in Tier 3 serve to further add additional qualitative dimensions for considering whether to build or buy.

TIER 3

Tier 3 is comprised of several decision points, with eight of them represented in Figure 31. Your company is free to augment this list with what is important to your organization as decision points in choosing "build" versus "buy." These decision points offer binary choices. On the left of the decision points is a "BUILD" decision, on the right of the decision points is a "BUY" decision. An answer of "YES" on any one of these decision points shifts your answer for that particular decision to the left, towards a "build" decision;

whereas a "NO" response shifts that decision point to the right, tilting to a "buy" decision.

The easiest way to use this tier is to denote each decision with a checkbox or some demarcation on the left of the decision or the right of the decision depending on your preference. Once complete, tally the number of decision points that decided to the left (a "build" decision) and the number of decision points that decided to the right (a "buy" decision). Whichever side has a majority of the decision points, informs which model should be pursued for the organization. The ***qualitative*** aspect of this framework comes in when there is an equal tie between both, the "build" and the "buy" models. This is part art and part science, and in the event of an even split, the decision should be up to the judgment of the organization's leadership team.

This tier offers eight decision points. It is strongly recommended that if using this framework, these basic eight decision points be left intact, and any additional ones be additive to the decision point list. The eight basic decision points (with binary "yes" or "no" responses for each as the only accepted answer), listed in no particular order of importance, are:

1. Ability for the organization to provide hosting and infrastructure.

2. Ability of the organization to provide ongoing support, maintenance, and regular updates.
3. Organization has mastery of pertinent data.
4. Organization has requisite skills and talent in-house.
5. Organization has a reasonably high data quality.
6. Organization requires a high degree of customization.
7. Organization needs absolute control over AI and ML models.
8. Ability of the organization to scale out commensurate to vision, strategy, and goals.

B. Tools for the "Build" versus "Buy" Decision Process – The Quantitative Decision Scorecard

The quantitative decision scorecard presents a simple mathematical manner to arrive at a build or buy recommendation. As depicted in Figure 32 below, this scorecard is built on identifying several measures and assigning these measures a weighted score.

This scorecard identifies twenty measures, but the framework is flexible enough to accommodate as many additional measures that a company deems is pertinent to their needs, as long as the rubric for arriving at a decision, as depicted later in Figure 33, is appropriately calibrated.

Domain	Measure	Score (1 – 3)			Weight (1 – 5)					Weighted Score (1 – 15)
Process	We have a Long-Term Vision for our Enterprise AI Program	1	2	3	1	2	3	4	5	
Process	We have Organizational Readiness across Enterprise	1	2	3	1	2	3	4	5	
Process	We have an Existing / Mature Enterprise Data Strategy and Governance Program	1	2	3	1	2	3	4	5	
Process	We have completed an Assessment by Technology Consulting Firm or Provider	1	2	3	1	2	3	4	5	
Process	We have completed an Assessment and Recommendations Exist by Risk Management Firm/s	1	2	3	1	2	3	4	5	
Process	We have Clearly Defined ROI / Business Value Aligned to Goals	1	2	3	1	2	3	4	5	
Process	We have a Need for Control over the Platform Roadmap	1	2	3	1	2	3	4	5	
Process	We Operate Under Time-to-Market Constraints	1	2	3	1	2	3	4	5	
Process	We are Ahead of Regulatory and Compliance developments	1	2	3	1	2	3	4	5	
Process	We Require / Prefer a High Degree of Customization / Bespoke Solutions	1	2	3	1	2	3	4	5	
People	We have Clearly Defined Roles and Responsibilities Across the Enterprise for AI	1	2	3	1	2	3	4	6	
People	We have Support and Maintenance Funding, Infrastructure, Processes, and Talent	1	2	3	1	2	3	4	5	
People	We have Talent Available in the Market / Within the Enterprise	1	2	3	1	2	3	4	5	
People	We have Skills Available Within the Enterprise	1	2	3	1	2	3	4	5	
People	We have a Chief Data Officer / Chief Data Analytics Officer / A Central Leadership Role Focused on Data	1	2	3	1	2	3	4	5	
Technology	We have Relative Technological Maturity	1	2	3	1	2	3	4	5	
Technology	We have all Pertinent Data in One Place / Federated / Mastered	1	2	3	1	2	3	4	5	
Technology	We have Reasonably High Data Quality	1	2	3	1	2	3	4	5	
Technology	We have Defined Plans to Scale Out our Technology	1	2	3	1	2	3	4	5	
Technology	We have the Ability and Capacity for Technology Hosting and Relative Infrastructural Maturity	1	2	3	1	2	3	4	5	
									TOTAL WEIGHTED SCORE:	

Figure 32: The Quantitative Build versus Buy Decision Scorecard

The twenty measures chosen for this scorecard have been selected to serve any company across any industry. It is highly recommended that these measures be expanded or added onto, but not removed or fundamentally altered.

These twenty measures that can be scored on a three-point scale, and assigned weights on a five-point scale are:

1. We have a Long-Term Vision for our Enterprise AI Program.
2. We have Organizational Readiness across Enterprise.
3. We have an Existing / Mature Enterprise Data Strategy and Governance Program.
4. We have completed an Assessment by Technology Consulting Firm or Provider.

5. We have completed an Assessment and Recommendations Exist by Risk Management Firm/s.
6. We have Clearly Defined ROI / Business Value Aligned to Goals.
7. We have a Need for Control over the Platform Roadmap.
8. We Operate Under Time-to-Market Constraints.
9. We are Ahead of Regulatory and Compliance developments.
10. We Require / Prefer a High Degree of Customization / Bespoke Solutions.
11. We have Clearly Defined Roles and Responsibilities Across the Enterprise for AI.
12. We have Support and Maintenance Funding, Infrastructure, Processes, and Talent.
13. We have Talent Available in the Market / Within the Enterprise.
14. We have Skills Available Within the Enterprise.
15. We have a Chief Data Officer / Chief Data Analytics Officer / A Central Leadership Role Focused on Data.
16. We have Relative Technological Maturity.
17. We have all Pertinent Data in One Place / Federated / Mastered.
18. We have Reasonably High Data Quality.
19. We have Defined Plans to Scale Out our Technology.
20. We have the Ability and Capacity for Technology Hosting and Relative Infrastructural Maturity.

Calculating the "Build" versus "Buy" Score

This scorecard helps to arrive at a weighted score that is based on an identified list of measures. Scores are assigned to these measures on a three-point scale. Weights are assigned to these measures on a five-point scale. Weights to a particular measure denote how important that particular measure is to an organization. A weighted score per measure is calculated by multiplying the raw score for a measure by its assigned weight. This weighted score helps in making a build versus buy decision when measured against the scale presented in Figure 33.

Here are the steps that one can use to arrive at a score, measure this calculated score against the scale in Figure 33, and use this as a quantitative means to decide on a "build" versus "buy" model:

1. Start with the twenty measures as identified. This list can be additive in nature, that is, additional measures can be introduced to this scorecard, if the arithmetical integrity of the scorecard is maintained. Additional measures will mean additional weighted measures, and as such the scale depicted in Figure 33 will need to change commensurate to that.

2. Score each measure from 1 to 3 depending on the level of maturity of that particular measure within your organization, 1 being the least mature and 3 being the most mature.

3. Weigh each measure from 1 to 5 depending on the level of importance of that particular measure within your organization, 1 being the least important and 5 being the most important.

4. Multiply the Score for the measure with the Weight of that measure to arrive at a Weighted Score. A Weighted Score for each measure will range from a minimum of 1 to a maximum of 15.

5. Repeat this calculation for each of the twenty measures.

6. Sum the Weighted Score for each of the twenty measures. This is the "build" versus "buy" Total Weighted Score. This Total Weighted Score can range from a minimum score of 20 to a maximum score of 300.

7. Evaluate this score against the rubric depicted in Figure 33, the decision scorecard scale. With a midpoint of 160, the closer the score is to the maximum of 300 points is an indication that the organization is able to undertake building and maintaining AI models ("Build"). Any score less than 160 points indicates that the organization might be better served in pursuing a partnership with a technology provider to avail itself of a full-service model ("Buy").

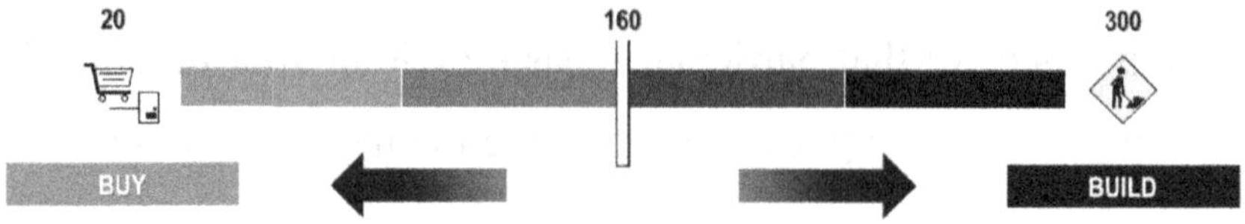

Figure 33: The Quantitative Build versus Buy Decision Scorecard Scale

Chapter Nineteen: Ten Enterprise-Level Best Practices – Part 3

Enterprise Best Practice 3 – "Quality Assurance and Quality Control Throughout" - explores the critical importance of Quality Assurance (QA) and Quality Control (QC) in your enterprise AI program. It is alluring to consider that a "buy" model effectively outsources the responsibilities of QA and QC to the third-party vendor/provider, but that would be an egregiously incorrect assumption. Regardless of whether a company chooses to "build" by developing AI models in-house, or "buy," by procuring services from third-party vendors, QA and QC are indispensable aspects of your enterprise AI program. QA and QC within your enterprise AI program underpin the success, reliability, and ethical deployment of AI solutions.

Whether your firm pursues a "build" approach or a "buy" approach, it is important to keep in mind that the ultimate responsibility for any output of an AI engine that is conducted on your behalf, is yours. Therefore, even if your organization decides to fully outsource the enterprise AI model building and maintenance to a vendor, it does not diminish your firm's responsibilities towards QA and QC, and in some sense, might actually need additional focus and oversight.

Enterprise Best Practice 3: Quality Assurance and Quality Control Throughout

As discussed earlier in this book, QA and QC have different connotations. Although they are frequently used together, and just as frequently used synonymously, they are not interchangeable. QA applies to the *process* of building an AI product. QA seeks to address the question if the AI product's components, and AI's product building processes, meet stated requirements. QC applies to the finished AI product and seeks to address the question of whether the AI product meets stated requirements.

The distinction between QA and QC can be visualized with two simple examples. Figures 34 and 35 represent two simple supervised and unsupervised machine learning examples respectively.

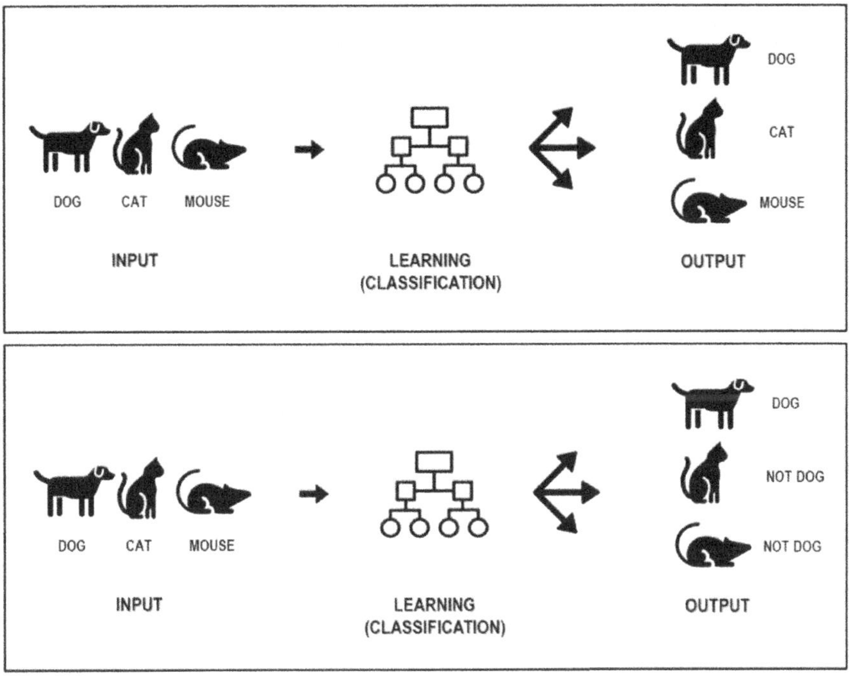

Figure 34: Simple Supervised Machine Learning Example

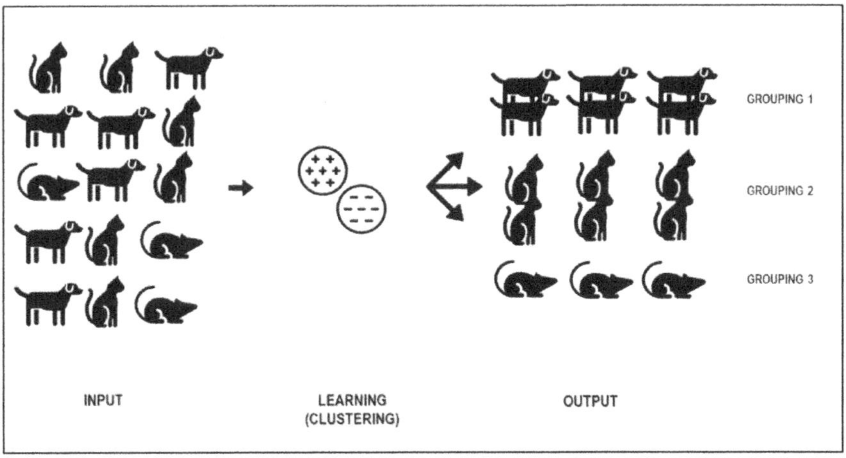

Figure 35: Simple Unsupervised Machine Learning Example

Figures 36 and 37 represent the same examples of supervised and unsupervised machine learning with QA and QC identified (QA and QC enclosed within black-bordered boxes). Note that these simple examples can be considered as an AI system.

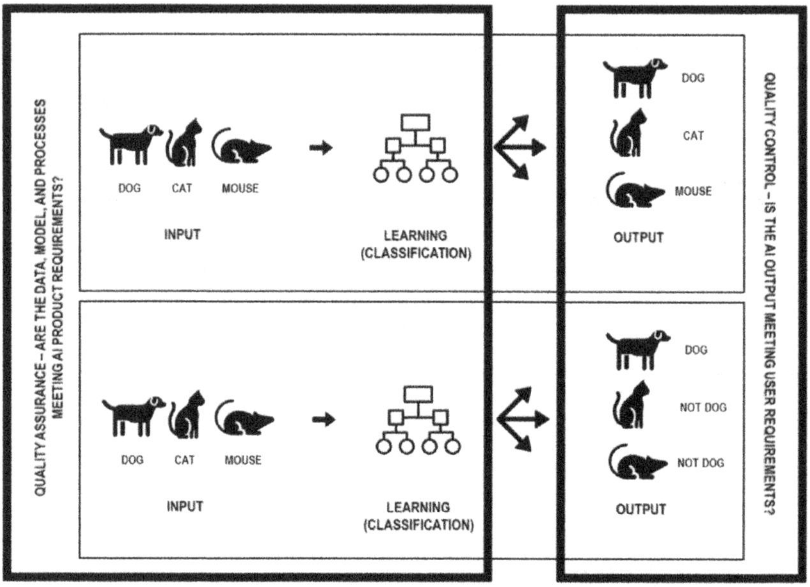

Figure 36: QA and QC in Supervised Machine Learning

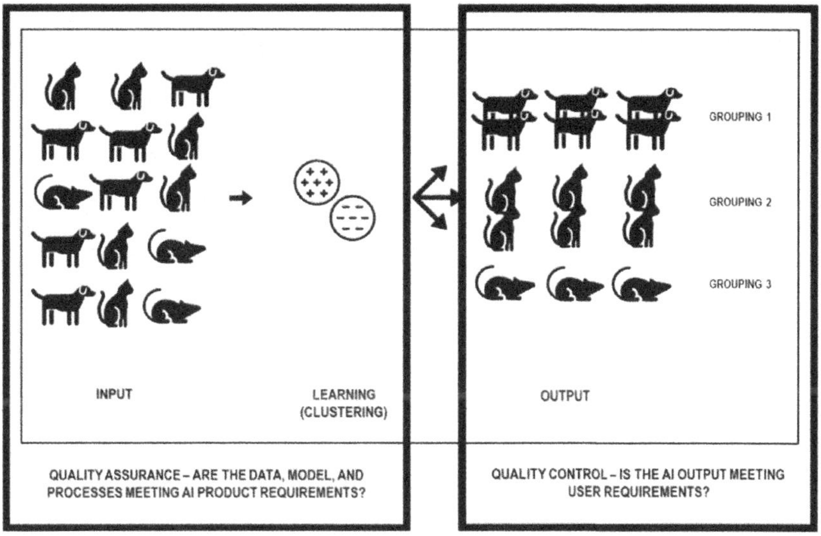

Figure 37: QA and QC in Unsupervised Machine Learning

Testing within the context of an AI system is predicated on the combination of two facets – testing of the data, and testing of the models. There are differences around who is responsible for both of these facets depending on who owns the AI models, but testing of the data should happen at both endpoints – the source as well as the destination (in the case of the "build" model). Data and model testing should be a key facet of an organization's AI program. If the "build" model is chosen, it is vital for companies to have an understanding of the provenance of external data being provided by technology providers. It is just as important for the firm to be able to track the lineage of this external data as it becomes internal to the organization's AI/ML ecosystem.

Implicitly, there should be clear, defined, and institutionalized measures of success around data used for AI when selecting a particular data source. Both quality control and quality assurance need to be an integral part of data testing. Organizations that have invested in data strategy and governance, and data quality, are the ones who generally have robust data testing practices. Data quality is a multi-dimensioned and multi-faceted aspect of this process and involves more than testing and validation of these external data sources. A key part of ensuring data quality is also data literacy, which is needed across an organization.

Companies that use the "buy" model generally look at models owned by technology providers as "black boxes." Companies send their business rules and associated data to their technology provider, and the provider in turn returns a recommendation/AI output. There is inherent trust placed by companies with these established and mature technology providers. A majority of the time firms are unaware of which, or how many, external data sources a technology provider uses for rendering an output - a decision or a recommendation. However, certain emerging regulations might necessitate companies to have a view into a manifest of data sources being used by the technology provider and how this data is being used. Without divulging their proprietary AI models, technology providers would serve themselves well by preparing to showcase at least some of their primary non-traditional data sources. Note that some emerging regulations seek to confer accountability - and this accountability will reside with the organization who owns the AI decision. In this instance, explainability will be important and a company - in order to demonstrate accountability – would benefit by understanding the data sources being ingested into their technology provider's proprietary AI/ML models.

Testing of the AI/ML models should happen depending on who owns the model – the organization or the technology provider. In either "build" or "buy" scenario, companies place intense focus on performing random sampling and audits of AI outputs by processing decisions by both - analog/digital, and their associated AI means. Companies who choose the "build" route place due emphasis on testing their AI/ML models in order to ensure they test for methodological issues, statistical soundness, model degradation, model explainability, and testing to address ethical concerns. According to a research paper for the financial services industry, analytics teams must be able to "demonstrate to internal stakeholders, customers, and regulators that their statistical methods are sound, that their models continue to perform well long after implementation, and that their results are reasonable and explainable. If their models do not follow accepted methodologies, or their performance degrades over time, then business partners will not trust the results, and regulators may not allow their use. If they are not able to interpret the results and explain why the models make the decisions they make, then they will not be able to gain full business value from their insights. For these reasons, analytics teams need to ensure that their statistical methodologies stay up to date. They also need to update their models regularly and be able to explain any changes in the results" (Kamath, Clark, Purushothaman, & Sakthivel, 2021).

Some firms prefer to conduct their methodological reviews internally versus inviting external review, while others invite external reviewers. Those who choose to keep testing controlled internally do so over concerns of needing to protect their data and proprietary models, and concerns that external reviewers might not understand their business model and custom business rules. It is important for companies to maintain statistical soundness. The same paper cited above states that - "To maintain statistical soundness in their analytics practice, companies generally rely on peer reviews by their internal analytics team, both during and after model construction and implementation. They strive to maintain transparency into the data sources and statistical methods they use during construction" (Kamath, Clark, Purushothaman, & Sakthivel, 2021).

As discussed earlier in this book, one of the key considerations of a "build" strategy is that AI/ML models need constant maintenance and upkeep to prevent degradation and model drift. The study on AI practices in the life insurance industry noted above continues - "Analytics teams need to ensure that their models do not degrade in performance over time. Several factors cause models to become less accurate as more time passes from their initial implementation. For example, the underlying data might not update regularly, making it impossible for a model to keep up with subsequent changes in the phenomenon it is trying to describe. Another troublesome possibility is that new data coming into a model can look different from the data used to train it, with some variables measured differently or missing entirely. This can break a model and generate misleading results" (Kamath, Clark, Purushothaman, & Sakthivel, 2021).

A frequently cited concern with firms regardless of industry or sector is the need to ensure models are transparent, explainable, and free from inadvertent bias or proxy discrimination. Ethical concerns are top of mind for organizations when it comes to testing. This is something that continues to pose a challenge to prove, since you cannot prove the absence of biased data, without explicitly introducing data that could simulate bias in training data for the models. Companies, including technology providers, cite ethical concerns to be at the top of how and why they test their models, with regular internal reviews being conducted that include their legal and compliance teams. Explainability is a key facet of testing, as highlighted by a research study in the financial services industry that states, "Analytics teams have a strong interest in being able to explain the results of their models. Even the most sophisticated model does not have any value unless the analyst can explain to business partners how it should be applied. An unexplainable model may also cause regulatory problems or ultimately become less likely to be adopted. To improve explainability, analytics teams tend to rely on documenting their data sources and variables thoroughly," concluding with, "Overall, analytics teams go to great lengths to mitigate methodological challenges that might diminish the accuracy of their models or impede their adoption. Peer reviews, continuous performance monitoring, periodic refreshes, and transparent documentation ensure their models are statistically sound at deployment, stay effective as time passes, and are explainable to non-technical stakeholders" (Kamath, Clark, Purushothaman, & Sakthivel, 2021).

The Importance of Focus on QA and QC

Robust QA and QC processes are foundational to ensure AI output precision, reliability, compliance with ethical standards, and continual improvement of AI solutions.

QA and QC in the AI development process offers significant benefits to any organization. Some of these benefits are like any other software or technology product, but with AI, there is a much deeper need for an intense and intentional focus on quality. QA and QC processes help to safeguard your organization against risks, and enhance the value derived from your enterprise AI program. Companies must prioritize QA and QC efforts as table stakes - foundational elements in the AI journey. AI systems can render decisions in nanoseconds, ingesting enormous amounts of data. The central premise of Explainable AI (XAI) is that a human should be able to clearly understand how and why an AI system arrived at the decision (output) that it did. QA and QC are key parts of ensuring that there is clarity and transparency around the holistic AI system. There are a few shared commonalities in the value of focusing on QA and QC across the AI value chain between the "build" approach and the "buy" approach, with some benefits unique to each.

Considerations for QA and QC Regardless of "Build" vs. "Buy" Approaches

Regardless of whether companies choose to go the "build" route, or the "buy" route, here are three considerations for ensuring QA and QC are appropriately prioritized in the AI development process.

1. Data Governance and Data Privacy: Both approaches, "build" and "buy," require stringent QA and QC measures for data governance and data privacy compliance. Ensuring adherence to privacy regulations, data security protocols, and ethical guidelines is crucial, regardless of the origin of AI models or services.

2. Continual Measurement and Continual Improvement: As reviewed in Basic Principle 4, QA and QC efforts extend well

beyond the initial AI implementation phase. Continual monitoring, evaluation, and feedback mechanisms are imperative to identify irregularities, biases within the model output, or performance degradation of AI models over time. This helps in continual improvement and refinement of AI systems, and is a crucial facet to ensure that these systems remain effective and ethical.

3. Explainability: As we will cover later in this book in the chapter on Explainable AI (XAI), QA and QC processes play a fundamental role in risk mitigation and addressing ethical considerations associated with AI adoption. The ability to proactively detect, diagnose, identify, and mitigate risks, including but not limited to concerns regarding bias, fairness, transparency, and security exposures, are vital to ensure responsible AI use.

Considerations for QA and QC with the "Build" Approach

The following aspects are notable considerations for a "build" approach, with regards to QA and QC within the AI development process (in addition to the considerations that are common across both approaches).

1. AI Algorithm Code Quality: Like with any software product, QA and QC, with a "build" approach have to focus on quality of

the code that your company's AI algorithms are built upon. The onus is on the company to ensure that your AI algorithms and the underlying infrastructure adheres to industry best practices, the highest cybersecurity standards, and aligns with regulatory requirements. Ongoing maintenance and updates, as with any other software product, are necessary to ensure that your AI algorithms are kept updated, performant, and secure.

2. Precision and Performance of AI Models: QA within the build approach is predicated on ensuring the precision, performance, scalability, and dependability of your AI models. As is the case with any technology initiative that your firm undertakes, thorough testing, validation, and verification processes are foundational in order to ensure that your AI models produce accurate and reliable outputs, as expected by your business requirements. QA also ensures that your AI models are corrected regularly to mitigate model drift. Institutionalizing QA in your AI development process is necessary to ensure robustness of your AI ecosystem and mitigate model performance degradation over time.

3. AI Model Data Integrity: As we have seen in Basic Principle 4, the continual monitoring and validation of data inputs into your AI models help in maintaining the quality of AI models over time. In AI development, QA and QC processes focus on data integrity. This ensures that the data set used to train and build AI

models is accurate, clean, reliable, fit-for-use, representative of the real world, and devoid of biases.

Considerations for QA and QC with the "Buy" Approach

The following aspects are significant considerations for a "buy" approach, with regards to QA and QC within the AI development process (in addition to the considerations that are common across both approaches).

1. Vendor Vetting: QA and QC efforts in the "buy" model must be focused on a thorough and rigorous vetting of the vendor/s being selected. A vendor assessment, and conducting due diligence on the vendor, from their cybersecurity posture to their reputation within your industry, is critical. Companies should evaluate any prospective vendor's record, their depth and breadth of expertise, examine referrals, and thoroughly evaluate the quality of the vendor's AI solutions. Even if your firm has a long-standing relationship with a vendor, downplaying or bypassing thorough vetting will not yield success. In addition to ensuring that a vendor can be a good partner, your firm will benefit from conducting a thorough assessment on the vendor's QA and QC practices, how they handle their data, how lucidly they can integrate with your company's ecosystem, how easily they can integrate with your enterprise processes, the vendor's command on the evolving AI regulatory and compliance landscape, and the

vendor's adherence to the highest possible ethical standards. Ideally, your vendor/s should hold themselves to a much higher and stringent ethical standard than you believe is adequate.

2. Integrations: Any company can partner with any vendor to avail themselves of their services for a "buy" approach. However, the real challenge will be in the myriad of integrations between the vendor's systems and your organization's ecosystem. These integrations are across technology, processes, and people (how will teams operate with each other). The people, process, technology triad will be deterministic on how effective you and your vendor partner can develop a shared understanding of sound QA and QC practices. Thorough testing is mandatory when partnering with a vendor for procuring AI solutions to ensure compatibility and consistency between your firm and the vendor.

Risks of Inadequate QA and QC

Poorly defined and/or poorly executed QA and QC can lead to a plethora of risks and challenges. Inadequate QA and QC can hamper your enterprise AI strategy, and compromise the effectiveness, reliability, and sustainability of your AI program. Inadequate QA and QC poses significant risks and challenges across both "build" and "buy" approaches. These challenges include the possibility of AI models that are inaccurate, security exposures, compliance failures, squandered resources, overreliance on vendors, and overall, pose a risk to the sanctity of your AI program. Figure 38 provides a high-level look at some of the risks and challenges resulting from poor QC and QC practices across both "build" and "buy" approaches.

BUILD	BUY
Questionable Model Accuracy Without defined QA and QC practices, your company's AI models might lack accuracy. This will lead to unreliable outputs, including misleading predictions, and/or decisions predicated on incorrect data or flawed algorithms.	**Overreliance on Vendors** Without thorough assessment and due diligence of potential AI vendor/s, the reliance on third-party AI services may expose your organization to being overly dependent on vendors. If the vendor/s have quality issues in their AI deliverables, these will certainly become your issues.
Potential for Biased Data Inadequate QA and QC around your data can result in biased or incomplete training datasets for your company's AI models. This can lead to the development of biased AI models that reflect societal or systemic biases.	**Vendor Integration** Inadequate integration testing and benchmarking may lead to compatibility issues or underperformance of the purchased AI service within your company's existing ecosystem. It is relatively easy to partner with a vendor, but integrating their offerings into your ecosystem can be challenging.
Cybersecurity Exposures Insufficient code review and cybersecurity QA and QC may leave your AI systems vulnerable to cybersecurity vulnerabilities. This can make your AI systems susceptible to data breaches, make them a potential threat vector, and expose you to the risk of compromising sensitive information.	**Data Privacy and Information Security** Despite all appropriate vetting by your firm, the lack of proper data validation and security checks within a vendor's ecosystem can result in data privacy issues, data breaches, or security vulnerabilities in the third-party AI service. Any missteps by a vendor performing AI services, with your data, on your behalf, might be considered as your responsibility.
Compliance and Ethical Concerns Lack of QA and QC, and adequate rigor around compliance checks and ethical considerations, can lead to being non-compliant with global data privacy regulations, or ethical AI guidelines, risking legal repercussions or significant reputational damage.	**Service Level Agreements (SLAs)** Inadequate monitoring and failure to establish clear SLAs might lead to service-level agreement violations or poor-quality services from vendors.
Time, Cost, Labor, Material Investments Inadequate QA and QC could result in the creation and subsequent maintenance of flawed AI models. This will certainly result in squandered time, cost, labor, and materials. The sunk costs will also pose challenges that would require rework, management of belied stakeholder expectations, implementation delays, and time-to-market challenges.	**Alignment with your Enterprise AI Strategy** Partnering with a vendor means that you will cede some of your enterprise AI vision to what the vendor has in mind for their AI product roadmap. Customizations, while often undesirable, are often necessary to further your unique strategy. Decoupling from vendors and switching is also notoriously challenging and can be an expensive proposition.

Figure 38: Risks and Challenges of Inadequate QA and QC in "Build" and "Buy" Approaches

Addressing these risks requires robust QA and QC practices to be established within your firm. These practices must be custom to each approach in order to mitigate the potential risks inherent to each, and in order to ensure the sustained success of your enterprise AI program.

Common QA and QC Techniques

As we have seen, QA and QC are essential to ensure the accuracy, reliability, optimal performance, and long-term success of your enterprise AI program. Regardless of which approach your company chooses – a "build" approach, or a "buy" approach – there are some common QA and QC techniques for your enterprise to consider incorporating into your AI program. Figure 39 provides an overview of common QA and QC methods specific to each approach.

BUILD	BUY
Testing and Validation 1. Unit Testing: Test individual components or functions of AI models to ensure they function as intended. 2. Integration Testing: Verify that different modules or components within the AI system work together seamlessly. 3. Validation Testing: Assess whether the AI model meets the specified requirements and delivers accurate outputs.	**Vendor Assessment and Due Diligence** 1. Vendor Evaluation: Assess the prospective vendor's expertise, reputation, reliability, and track record in delivering AI services. 2. QA Process Evaluation: Review the prospective vendor's QA processes, certifications, and standards to ensure quality control measures align with your company's requirements.
Data Quality Assurance 1. Data Preprocessing: Clean and preprocess the raw data to remove noise, handle missing values, and standardize data formats. 2. Bias Detection: Employ statistical techniques and fairness metrics to identify and mitigate biases in training data.	**Integration Testing and Performance** 1. Integration Testing: Validate the compatibility and interoperability of the vendor AI solution with existing systems. 2. Performance Benchmarking: Test the purchased AI service against your predefined benchmarks to ensure it meets your expected performance levels.
Code Review and Documentation 1. Code Quality Checks: Conduct code reviews to ensure adherence to coding standards, optimize performance, and identify potential vulnerabilities. 2. Documentation: Ensure comprehensive and accessible documentation exists for AI models, in order to help in understanding, maintenance, and future improvements.	**Data Validation and Security Compliance** 1. Data Validation: Verify data integrity and privacy safeguards are implemented by the vendor/s to ensure compliance with your company's standards. 2. Security Assessment: Evaluate vendor's AI security measures, encryption protocols, and access controls to protect sensitive data.
Security and Compliance 1. Security Audits: Evaluate the AI system's vulnerability to cyber threats and implement security protocols to safeguard against potential risks. 2. Compliance Assurance: Ensure compliance with data privacy regulations and ethical guidelines, such as CCPA/CPRA/PIPL/GDPR or ethical AI frameworks (example: EU).	**Service-Level Agreements (SLAs)** 1. SLA Specification: Define clear SLAs with the vendor/s, outlining performance expectations, uptime, support, and resolution times. 2. Continual Monitoring: Implement monitoring tools to continuously track the performance of the vendor's AI service/s and adherence to established SLAs.
Regular Audits and Feedback Loop Process 1. Periodic Audits: Conduct regular audits of your AI models and outputs to ensure ongoing compliance, security, and quality. 2. Feedback Mechanisms: Establish internal feedback loops to provide insights and suggestions for improvements, fostering continual enhancement of your AI models.	**Regular Audits and Feedback Loop Process** 1. Periodic Audits: Conduct regular audits of the vendor/s AI service/s, and your integrations with them, to ensure ongoing compliance, security, and quality. 2. Feedback Mechanisms: Establish feedback loops to provide insights and suggestions for improvements to the vendor/s, fostering continual enhancement of the AI service/s.

Figure 39: Common QA and QC Methods for "Build" and "Buy" Approaches

The "build" and "buy" approaches for AI require distinct, yet overlapping, QA and QC methods. Regardless of the model you ultimately choose, your AI program should be grounded on institutionalizing QA and QC across the value chain to ensure the reliability, accuracy, security, scalability, and the ethical, and long-term, sustained success of your AI program. Enterprise Best Practice 4 – AI Body of Practice (AI BoP) helps you gauge the maturity of your AI program, and implicitly, the QA and QC maturity inherent to your AI practice.

Chapter Twenty: Ten Enterprise-Level Best Practices – Part 4

Most industries that have been slower in digitally transforming themselves, have also struggled to establish a culture where data is treated as an enterprise asset. Driven by the investments in digital transformations, these industries have recognized the vitality of data to enable these large-scale digital transformations. Data, and all aspects of managing data - including data strategy, data governance, and data management - now feature as a key strategic investment across companies of all sizes. Having treated data as a byproduct of systems for decades, rather than a product that can be monetized and serve as a competitive advantage, the implementation of data strategy and governance programs across these industries have yielded mixed results. Not every company has demonstrated sustained success with their programs, and even fewer have been able to inspire a true data-driven culture. The organizations that have however been able to make sustained investments in their data strategy and governance programs, and been able to derive value from them, have been more successful in their digital transformations. These organizations have been able to make measurable progress in inspiring a data-driven cultural mindset throughout their firms.

While data strategy and governance programs were still maturing across these industries, AI entered the picture and heaped additional pressure on organizations to have a sound and robust plan on how to effectively manage their data. The rapid expansion of AI across these industries has provided additional impetus for companies to not only have a well-defined data strategy, but also invest in their fledgling AI bodies of practice. The simple equation that we established earlier in this book, that companies would do well to keep in mind is, "Good AI + Bad Data = Terrible AI." Enterprise Best Practice 4 recommends that firms invest in development of an AI Body of Practice, the structure of which shall be delved into in further detail in this chapter.

Enterprise Best Practice 4: AI Body of Practice {KEY RECOMMENDATION}

The AI Body of Practice (AIBoP) is a benchmarking instrument. The AIBoP is intentionally stylized similar to the concept of a Capabilities Maturity Model (CMM) that is leveraged across most companies. As with most CMM concepts, this AI Body of Practice should be enabled by multiple facets, from skilled and trained employees, appropriate enabling technologies, repeatable processes, education and training, robust documentation, standards and practices, good governance models, communication, and collaboration across the value chain, etc. The AI Body of Practice provides a mechanism for organizations to benchmark their AI and

ML maturity in an effort to invest in the appropriate areas to continue achieving increasing levels of operational maturity.

AI Body of Practice – Maturity Model

The Capabilities Maturity Model (CMM) is a popular framework within organizations to measure their organizational maturity on a specific topic or domain. Over the years, variations of the CMM have been applied to data strategy, data governance (Gupta & Cannon, 2020), and data management (Sweden, 2009), as well as general IT practices, IT Risk Management (Carcary, 2013), etc. The AI Body of Practice for AI model maturity, inspired by the Capabilities Maturity Model, is presented in Figure 40.

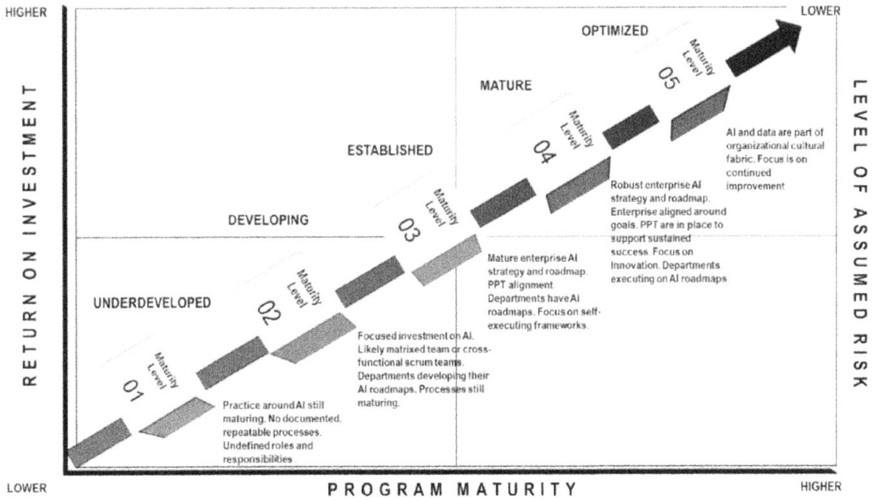

Figure 40: AI Body of Practice Maturity Model

This AIM Framework© recommends that organizations would do well to benchmark themselves on this AI Body of Practice maturity model as represented in Figure 40, with established goals to "level up" from their current benchmarked Maturity Level to the proceeding Maturity Level.

The maturity model depicts the maturity of an organization's AI program along the X-axis. The Y-axis represents the expected Return on Investment (ROI) of the firm's enterprise AI program. The Z-axis represents the level of assumed risk inherent in these programs. It is evident from this maturity model, that the more undeveloped an organization's AI practices are, they will yield a lower return on investment, and assume a higher level of risk.

This maturity model, the AI Body of Practice, benchmarks an organizations AI maturity across five levels:

- Level 01 - Underdeveloped
- Level 02 - Developing
- Level 03 - Established
- Level 04 - Mature
- Level 05 - Optimized

The least mature of the AI benchmarks is Level 01, "Underdeveloped," while the most mature is Level 05, "Optimized." Each of these maturity levels are explored a bit further in the following section.

AI Body of Practice - Maturity Model Levels

The following are the five AI maturity model levels. Just like with the CMM (or any derivation of it), leveraging a maturity model is part art and part science. There are some tangible measures you can employ to slot your organization into one of these five maturity levels, but more often than not, your organization's positioning into one of these five levels will come from how you "feel" about where your firm is (intangible), in combination with specific benchmarks (tangible). You know your organization the best, and instinctually you will know exactly what level your firm can be slotted into at the present moment. Based on your enterprise AI strategy – or at the least, your corporate vision for AI – you can then extrapolate at what level you aspire your firm to be, on a year-over-year basis.

Maturity Level 01: Underdeveloped

This is the lowest level of AI and ML program maturity. It presents the highest level of assumed risk and the lowest return on investment. Organizations at maturity level 1 have a practice that is still developing around AI, however, they lack the rigor around documentation and repeatable processes to be able to scale out this practice. Roles and responsibilities within organizations at this maturity level remain undefined. The progression of AI in organizations at this maturity level is generally unplanned and ad-hoc. Note that departmental AI strategies, in addition to an enterprise

AI strategy - is intended to alleviate the ad-hoc and unplanned proliferation of AI solutions within a company.

Maturity Level 02: Developing

This is the second level of benchmarked maturity that a company's AI practice can achieve. Organizations in this category have a focused investment in AI. They typically have matrixed organizations, or cross-functional scrum teams that are focused on building out prioritized projects that fall under their enterprise AI programs. Their departmental AI journeys might have a vision, however the strategy, roadmap, structure, and processes required to enable execution are still maturing. To scale out their AI programs, these organizations would benefit by investing in organizational structure, and ensuring that they are building scalable and sustainable operational processes.

Maturity Level 03: Established

Organizations in this maturity level benchmark enjoy relatively mature AI practices. They have well-defined and communicated strategies and roadmaps. They are effective in having been able to align their people, processes, and technology around their enterprise objectives, which are reflective of their vision for enterprise AI. Departments have departmental AI strategies and roadmaps that supplement and complement the enterprise AI strategy. These organizations can then focus on continual improvement and self-executing frameworks for achieving scale.

Maturity Level 04: Mature

Companies achieving this maturity level benchmark have robust enterprise AI strategies and programs. Their strategies serve as case studies for other organizations and the industry to emulate. The entire enterprise is aligned around their business objectives and goals with AI as a prominent enabler for achieving these goals. These types of organizations have people, processes and technology in place to support their programs, for sustained success, and accomplish the vision that they set for themselves. These organizations, with their AI programs (enterprise and departmental) running like "well-oiled machines," can invest heavily in innovation and innovative practices. The types of organizations at this maturity level are those that are going to help lead and transform their own industries and are considered as progressive and innovative leaders in their field.

Maturity Level 05: Optimized

This is the highest achievable benchmark in this maturity model framework. Organizations that have achieved a maturity level of 05 are successful in having mature and scaled AI practices, as well as high quality data, as an integral part of their organization's culture. These companies are innovative, are considered leaders in the field, and have been able to scale out their AI programs to every facet of their organization (enterprise and departmental). The focus for these firms is entirely on continual improvement, and they can afford to take measured and calculated risks, predominantly because their level of assumed risk from AI is significantly lower than the other four maturity levels. These organizations enjoy a significantly higher return on investment for their AI programs than others.

Level Setting

As mentioned earlier, calibrating your organization's AI maturity, and plotting this maturity to a level on the AI BoP Maturity Model is part art and part science. There are several options available to you in order to facilitate positioning your firm to one of the five levels. Figure 41 below recommends a few dimensions that you can measure quantitively and qualitatively. Figure 42 provides some mechanisms on how you can measure these dimensions - both quantitatively, as well as qualitatively.

	QUANTITATIVE	QUALITATIVE
Data Readiness	• Data Quality Index: Measure the accuracy, completeness, consistency, and relevance of available data. • Volume and Variety Assessment: Quantify the amount and diversity of data sources used for your AI models. • Accessibility Metrics: Analyze ease of access and availability of data across the organization.	• Data Governance Review: Assess policies, procedures, and controls governing data quality, security, and privacy. • Compliance Check: Evaluate adherence to regulatory standards (GDPR, PIPL, CCPA) and ethical data practices.
Strategic Alignment	• Alignment Score: Measure how well your enterprise AI strategy aligns with overarching business goals through KPIs. • Investment Ratio: Quantify the proportion of resources (financial, human capital) allocated to AI initiatives across the enterprise.	• Vision and Leadership Clarity: Assess the clarity and communication of the company's AI vision by the CEO and the C-suite, across enterprise leadership tiers. • Flexibility in Strategy: Evaluate adaptability to changing market conditions and technological advancements.
Technology Infrastructure	• Infrastructure Scalability Index: Measure your infrastructure's ability to scale with increasing AI demands. Note that for a "buy" model, measure your infrastructure's scalability commensurate to integration needs with your vendor/s. • Reliability Metrics: Analyze system uptime, performance, and failure rates.	• Integration Capability: Assess the ease of integrating new AI technologies with your existing systems. • Technological Flexibility: Evaluate the ability to adopt emerging AI tools and platforms.
Talent and Skills	• Skills Inventory: Quantify the number and proficiency levels of employees possessing AI-related skills. • Training Investment: Measure the hours invested in AI-specific training and skilling/reskilling/upskilling programs.	• Cross-functional Collaboration: Assess the level of collaboration among teams for AI-driven projects. • Innovation Culture: Evaluate your company's culture fostering experimentation and learning.
AI Applications	• Deployment Rate: Measure the speed and frequency of deploying AI applications. • Effectiveness Metrics: Assess the impact and success rates of deployed AI solutions through ROI and user feedback.	• Strategic Alignment: Analyze how AI applications align with your enterprise strategic business objectives. • Innovation and Adaptability: Assess your company's ability to innovate and adapt AI solutions to changing needs.
Ethical AI	• Compliance Score: Measure compliance with ethical frameworks and regulations, including Explainable AI (XAI). • Bias Detection Metrics: Assess the effectiveness of algorithms in detecting and mitigating biases.	• Ethical Framework Review: Evaluate policies and practices ensuring fairness, transparency, and accountability in AI (such as those outlined in the Explainable AI guidelines).
Risk Management	• Governance Index: Measure the level of formalized AI governance and risk management protocols. • Risk Identification Rate: Quantify the rate at which potential risks related to AI are identified and addressed.	• Policy Robustness: Assess the depth and effectiveness of policies related to AI governance and risk management. • Crisis Response Preparedness: Evaluate readiness in handling AI-related crises or failures.
Stakeholder Engagement	• Engagement Metrics: Measure customer satisfaction or stakeholder engagement impacted by AI initiatives. • Personalization Index: Quantify the level of personalization achieved through AI-driven interactions.	• Feedback Mechanism Effectiveness: Assess the effectiveness of your feedback loops for improving AI applications. • Communication Strategy Impact: Evaluate the impact of communication strategies regarding AI initiatives on stakeholders.

Figure 41: Potential Maturity Dimensions and Quantitative and Qualitative Measures

1. Scoring System	Develop a scoring system that works for your organization. Consider assigning scores or ratings based on a defined scale. Allocate weights to parameters based on their strategic significance. For example, assign scores (quantitative), or ratings (qualitative), to each dimension within the framework (on a scale of 1 to 5, with 5 indicating high maturity).
2. Data Collection	Employ various methods like surveys, interviews, assessments, audits, and data analysis tools to gather both quantitative and qualitative data. Ensure that this happens across the enterprise, especially with the departments engaged in prioritized AI projects.
3. Weighted Analysis	Assign weights to each parameter based on its significance to your organization's AI strategy and overall business objectives. Based on the scoring system you have decided to employ, assign scores or ratings based on a defined scale and allocate weights to parameters based on their strategic significance.
4. Benchmarking	Analyze the collected data against industry benchmarks, standards, best practices or previous assessments to benchmark and gain a deeper understanding of your company's AI maturity level.
5. Actionable Insights	Use the results of the assessment to conduct a SWOT analysis. Use the SWOT analysis to guide future AI strategy and improvements. Derive insights, identify strengths, weaknesses, opportunities, and threats, and formulate actionable recommendations to enhance your organization's AI maturity level.

Figure 42: Five Steps to Benchmarking to an AI Maturity Level

The dimensions as well as associated qualitative and quantitative measures are intended to seed thought. You might consider additional, or many other dimensions. Do not be hesitant to use the "gut feel" methodology as well when you are seeking to do a qualitative assessment of your company's AI maturity positioning. More than likely, this "gut feel" assessment is predicated on a depth of knowledge of your firm's maturity levels around people, process, and technology, as pertains to the adoption and implementation of AI. The "gut feel" approach, however, will itself be inadequate when reporting on your company's maturity level with senior-level stakeholders, especially in the C-suite. Your "gut feel" assessment, when reporting status or positioning to the C-suite, will need to be backed by some hard numbers. Note that it is likely that your company already has a methodology around what you measure, and how you measure for other CMM implementations. Some of the more common CMM derivations are around your software development practices, data practices, etc. The AI Body of Practice should be an overlay atop your existing measures, or derivations of these measures germane to AI. In other words, in seeking to calibrate your organization to one of the five levels, there is no need to "reinvent the wheel." Try to capitalize on the existing frameworks and infrastructure that your organization has in place, customizing for AI where appropriate.

To get a complete understanding of your organization's position on the AI maturity spectrum, you are at liberty to combine your existing internal assessments and benchmarks, network with peers across the industry (industry trade associations are an excellent source of facilitating this networking), attend conferences, and leverage research. It is recommended that you do not restrict yourself solely to research that is relevant to your industry. You should absolutely commence with industry-focused research, but in order to get a broader perspective - or leverage ideas outside of your own industry - deriving insights from research focused on other sectors, is a helpful exercise.

Enterprise Best Practice 5 examines the most fluid of all best practice focus areas – the rapidly evolving AI regulatory landscape.

Chapter Twenty-One: Ten Enterprise-Level Best Practices – Part 5

The Age of AI will be heralded as the most transformative technological advancement era in the 21st century. The accelerated proliferation of AI across industries has served as a catalyst for remarkable innovation and progress. However, this exponential growth of AI also raises profound ethical, societal, and legal concerns. Regulations and regulatory frameworks – especially in the digital age – are intended to protect us, the human consumers. Enshrining the digital rights of humans within regulatory frameworks has been the most effective way of ensuring digital identities are protected, and the right to privacy is followed. Regulation that balances the importance for continued innovation and experimentation, with some guardrails of what is permissible with AI is necessary to protect the digital rights of consumers (humans), enforce accountability, and to mitigate the potential risks posed by AI.

The need to strike a balance between innovation - with ethical considerations serving as a backdrop - there are a slew of proposed regulations and regulatory frameworks being developed across countries, blocks of countries, trading zones, sectors, industries, and even germane to specific use cases of AI within a particular industry. It is of paramount importance for companies to ensure that they are actively monitoring the regulatory landscape. If your enterprise AI strategy has active use cases that – at some point – are discouraged or even declared out of compliance with new and emerging regulations, as an organization, you must be ready to pivot and adjust in accordance with these frameworks.

Enterprise Best Practice 5: Monitoring the Regulatory Landscape {⚡ KEY FINDING/RECOMMENDATION}

The explosive growth of AI – and the fluidity of the AI field itself – has proven challenging for countries and regulatory bodies to develop regulations and frameworks to govern AI. There is no sign of this trend changing. Any guidance provided on AI regulation will likely be outdated in less than a year after the guidance was provided. This fluidity and rapid evolution of the regulatory environment is precisely why Enterprise Best Practice 5 recommends that organizations – regardless of industry, sector, size, or scale - pay special attention towards actively monitoring the regulatory environment.

The Importance of Regulation

The concept of Explainable AI (XAI) has been consistently raised throughout this book as pertains to transparency and trust with AI models. XAI, as has been mentioned, is foundational to build trust in

AI, and implicitly, ensure that AI is not treated as a "black box." It is vital for humans to understand why an AI system arrived at the decision that it did, and just as important for the human AI operator to be able to explain the AI's decision-making to other humans. Transparency in AI is important for the simple reason that without transparency in AI, we cannot entrust that an AI system is rendering decisions that align with ethics and that these decisions are not predicated on biased data, biased models, or both. This is where regulations are helpful. Regulations prescribe frameworks to allow organizations to confidently innovate, ideate, experiment, and implement sophisticated AI, knowing that following regulatory guidelines, they are unlikely to be susceptible to failing to comply with ethical standards. They protect the consumer as well as your organization. Trust is the bedrock of AI adoption, and regulation aims to instill confidence in humans – whether consumers, or stakeholders in a company - by enforcing transparency, explicability, mandating responsibilities, and attributing accountability in AI systems for failures or malpractices. This helps to foster trust among consumers and companies alike.

As AI continues to exert influence, if not outright dominance, over all aspects of our professional and personal lives, regulations will help to establish, and enforce, basic ethical standards. This will ensure that regardless of industry, or company, AI systems are congruent with basic human rights. Ethically aligned regulations encourage innovation and ongoing development within AI, seeking to maximize the transformation benefits that come from AI, while seeking to minimize any potential harm. If your organization operates across multiple countries around the world, AI regulatory frameworks can help provide your firm with a level playing field. Some level of AI regulatory standardization will help promote interoperability and help streamline compliance efforts for multinational companies. This will also make it easier for global companies to draft some common AI governing policies for their organizations, as well as make it easier for consumers to hold organizations accountable for noncompliance.

Regulations seek to ensure that AI systems are fair, transparent, free of bias, and that organizations operating these AI systems can be held accountable. They seek to protect consumer digital and data privacy, and prevent discriminatory practices – both inadvertent and intentional. Absent regulation (or self-regulation by ethically minded organizations), AI can perpetuate unintended consequences. Some of the resultant challenges of ill-defined regulatory frameworks are ethical, and illegal in nature, and can pose significant reputational hazards for your firm – for example, if a mortgage or a loan is denied by an AI system based on the applicant's race, gender, or gender identity. Some unintended consequences are more severe, and can have life-or-death implications – for example, if AI misdiagnoses a patient, or inadvertently causes a malfunction in life-support systems. There are very real safety standards concerning AI use - in addition to the ethical, legal, and reputational considerations - that regulatory frameworks help to mitigate potential risks with.

Regulated Industries

Monitoring the evolving regulatory landscape takes on special significance in highly regulated industries. Highly regulated industries, encompassing sectors like healthcare, finance, pharmaceuticals, aviation, and energy, operate within complex multi-dimensional frameworks of laws, standards, and protocols. Even if the industry that your company operates within is not a highly regulated industry, there are important lessons that your industry and your company can learn from observing advancements in the regulatory environment within these highly regulated industries. These industries can be considered as the gold standard when it comes to having institutionalized regulatory compliance, operating with the highest levels of ethics and transparency. With ethics and consumer protection having been ensconced in regulatory frameworks, these firms seek to exceed, if not meet, the highest levels of ethics and transparency as prescribed by the regulatory frameworks they are subject to. These are good examples to learn from, and base your own corporate AI Governance Policies upon. AI offers immense promise for these industries, but also necessitates a multi-faceted approach to regulation due to the complex implications on standards, safety, society, data privacy, ethics, and compliance.

Pharmaceutical development - and the biotechnology industry as a whole - is one such example of a highly regulated industry. The COVID-19 vaccine was developed in record time thanks to AI, and has likely saved millions of lives around the world. AI can greatly help in discovery of new and better life-saving drugs and such vaccines. In biotechnology, AI holds the promise of personalized treatment development. However, as with any life-and-death situations, regulation governing AI is vital to ensure the safety, efficacy, and ethical use of AI in developing drugs, vaccines, and personalized medicine to guarantee that patient care isn't compromised. Energy and Basic Utilities is another example of a highly regulated sector that is employing AI for a vast variety of tasks across the energy generation, storage, and distribution value chain. AI is used for predictive maintenance in combination with Internet of Things (IoT) sensors, grid management, fault tolerance and failover, demand management, emergency management, etc. AI regulation is crucial to ensure the stability (and protection) of energy grids. The transportation sector is another example of a highly regulated industry. With autonomous terrestrial and aerial vehicles (the latter being in pilots as of the first quarter of the 21st century), governed by AI systems, and, for the foreseeable future, having to cohabit the roads and skies with human-operated vehicles, regulation is essential to ensure the safety of passengers in autonomous vehicles, as well as the drivers of manned vehicles.

Thematically similar to Pharmaceuticals and Biotechnology, is the massive Healthcare Industry. A highly regulated and complex industry, AI is increasingly prominent in this field to help influence critical decisions in diagnosis, treatment, and patient care. Regulation is pivotal to ensuring that AI-driven medical devices, diagnostic tools, and treatment recommendations comply with rigorous safety standards and clinical validations. Regulatory organizations such as the Food and Drug Administration (FDA) have developed regulatory guidance frameworks for AI-based medical devices and software. These regulatory frameworks outline validation requirements, risk assessments, and continuous monitoring protocols in order to ensure the safety and effectiveness of AI-enabled healthcare solutions. AI is quite ubiquitous across the Finance Services sector, and is widely used across banking, lending, investment, and insurance industries. Some of the popular use cases of AI in these industries are for assessing and managing risk, customer service, customer engagement, customer knowledge, fraud detection, modernization, cost savings, and operational efficiencies. From fair lending, to maintaining stock market trading stability, to protecting customer privacy, regulatory frameworks across the Financial Services sector are vital. Within insurance specifically, we will explore an illustrative example of regulatory frameworks that seek to ensure that AI-driven underwriting is free of bias and proxy discrimination. There are AI use cases across the insurance value chain, regardless of type of insurance (life insurance, property and casualty, group and worksite, etc.), that would benefit from regulatory guidance.

As a guidepost on the unique nature of AI regulation across industries, and the importance of regulatory frameworks within those industries, Figure 43 explores the importance of AI regulation within Healthcare and Insurance. It is recommended that you follow a similar approach towards understanding what makes your particular sector and industry unique when it comes to AI regulation. If you are not in a regulated industry, you can draw inferences from what makes your specific industry unique, and draft your AI Governance Policies commensurate to this uniqueness. If you are in a regulated industry, and AI regulation in your industry is still evolving, your firm would benefit from following a similar exercise to base your own corporate policies on. Consider for a moment that absent overarching regulation, there is nothing to prevent your firm's AI Governance Policy from serving as a de facto template for your industry, or even serve to influence regulatory frameworks.

INDUSTRY-SPECIFIC CONSIDERATIONS AND IMPORTANCE OF REGULATION (HEALTHCARE AND INSURANCE)	
1. Existing Regulatory Frameworks: Healthcare and insurance deal with highly sensitive personal information. These regulations are based on data. These data and digital regulations make these industries subject to stringent existing regulations like Health Insurance Portability and Accountability Act (HIPAA) and General Data Protection Regulation (GDPR). AI's integration demands compliance with these regulations to protect customer confidentiality, ensuring secure data handling and preventing unauthorized access.	***4. Consumer Trust:*** Effective regulation helps to instill trust and confidence among consumers. Transparent AI that adheres to regulatory standards helps to foster trust by ensuring that decisions are fair, reliable, and aligned with ethical considerations.
2. Consumer Protection: AI applications directly impact customer well-being in healthcare. Regulatory oversight is vital to validate clinical efficacy, accuracy, and safety of AI-driven diagnostic engines, treatment recommendations, and patient care. Rigorous testing and validation procedures are essential to prevent misdiagnoses or treatment errors. In healthcare and insurance, AI regulatory frameworks help mitigate biases and ensure fairness in decision-making processes. Ethical considerations are paramount, because AI can influence life-altering decisions, such as medical diagnoses or insurance coverage determinations. Regulatory frameworks are required to establish guidelines for fair and equitable AI applications.	***5. Risk Mitigation:*** Adherence to regulatory frameworks mitigates legal risks and liabilities for companies in these industries. Compliance with established guidelines helps firms protect against potential litigations resulting from AI-related errors, biases, or breaches of data privacy.
3. Explainability: As the chapter on Explainable AI discusses, the "black-box" nature of AI algorithms poses challenges in explaining decision-making. In healthcare and insurance, transparency and explainability of AI-generated decisions are critical. Regulatory frameworks must ensure that AI-driven decisions are explainable to all stakeholders, fostering trust and accountability.	***6. Innovation:*** Well-crafted regulations strike a balance between innovation and compliance. This can nurture the development of responsible AI innovation in these industries. Clear guidelines from regulatory bodies helps to encourage investment in AI research and development, while ensuring that advancements benefit - and protect - both industry stakeholders and consumers.

Figure 43: Examples of Industry-specific Considerations and Importance of Regulatory Frameworks

Chapter Twenty-Two: Ten Enterprise-Level Best Practices – Part 6

Continuing the conversation around the regulatory landscape, this chapter shall explore an illustrative example of how regulatory frameworks are shaping up for one facet within one industry (life insurance).

This chapter shall also provide an overview of the two regulatory frameworks that warrant increased attention – the European Union (EU) AI Act, and President Biden's Executive Order on AI. An overview of guidelines from two governmental bodies is important – even if your organization does not fall under these jurisdictions, they are helpful frameworks to learn from, and base a safe AI strategy upon. Additionally,,of the two, the EU AI Act comes with penalties for noncompliance. However, it is foreseeable that the US EO serves as a template for regulatory agencies that govern specific US industries to create custom regulation on AI for the industries they oversee.

Illustrative Example – Regulatory Frameworks within the Life Insurance Industry

Expanding on the look at the healthcare and insurance industries, it is helpful to take a deeper look at a regulatory framework overview within one specific aspect of an industry. Presented here is an illustrative example of one of the AI use cases within the life insurance industry. It will be important to extrapolate lessons from this illustrative example for your own industry, and infer how you can co-opt some of the best practices that the cited regulatory frameworks propose for your own company. The life insurance industry is a highly regulated industry and can serve as a good illustrative example, regardless of the sector/industry that your company operates within. Even if your firm is not part of a highly regulated industry, the life insurance industry is a good example of an industry that has been historically slow to digitize. The industry has not, until recently, sought to capitalize on the wealth of its data assets, and it was not more than a decade to a dozen years ago that the industry began investing in digital transformations in earnest. The value proposition of the industry – to provide for financial security of millions of people around the world – also aligns with the core premise of any AI regulation – protecting the consumer.

Regulations in the United States, both at the Federal and State levels, are intended to protect consumers and implicitly, life insurance companies, and the industry at large. Regulations protect life insurance companies indirectly by developing and maintaining standards and by providing guidelines for the industry to follow. While some serving in the industry consider overregulation as inhibitory to innovation and growth, a dominant majority recognize the criticality of regulation to protect the consumer and the industry as a whole. An often-overlooked fact is that individuals who work in the industry and are subject to any regulatory compliance, are themselves consumers too, and as such benefit from the guardrails and parameters the regulations provide. The Federal Insurance Office (FIO) was created in 2010 in response to the financial crisis of 2008. Operating at the federal level, the FIO lacks any regulatory oversight and authority, but is responsible for monitoring insurance markets to ensure that any activities that could contribute to a financial crisis - like in 2008 - are intercepted. The FIO also monitors the coverage gap across America, with the hope that all Americans, especially those in underserved communities, are protected by access to affordable insurance.

Regulation in the life insurance industry is driven by individual states, which have significant autonomy in being able to set state-level regulations that apply to life insurance companies that operate and/or are domiciled in that state. The National Association of Insurance Commissioners (NAIC) in turn performs the critical function of serving as the central hub for all state insurance regulators. The NAIC is the body that recommends and sets industry best practices and standards, and therefore most states prescribe to the model regulations developed by this group. The NAIC regulatory recommendations provide the means to standardize how insurance companies operate across the United States. Evidenced by the lack of defined regulatory recommendations by the NAIC, or any other independent state regulatory body, life insurance companies have largely been operating in the absence of defined parameters for AI across the value chain. One such prominent use case of AI in life insurance is within the automated and accelerated underwriting space. Underwriting, enabled by AI, holds significant promise to streamline the process and bestow protection to an applicant in days versus four to six weeks after when a person applies for a life insurance policy. These faster turnaround times will be critical to serve the need of the next generation of digitally native consumers. AI-driven automated and accelerated underwriting grew so quickly and in such a short span of time within the industry, that even the state regulators, including the NAIC, have been measured in their ability to analyze, react, and respond with regulation that protects the consumer while ensuring more Americans are protected with life insurance coverage. The National Association of Insurance Commissioners (NAIC) established the Accelerated Underwriting Working Group (AUWG) in 2019 in order to evaluate the use of external data and data analytics in automated and accelerated underwriting. There are two other prominent regulatory guidelines that the industry has been closely watching. One of these guidelines is from the New York State Department of Financial Services. The New York State Department of Financial Services released a circular (Regalbuto, 2019) in 2019 that limits the use of external data, ML algorithms, and AI models that could cause proxy discrimination and inadvertent bias. This 2019 circular states that any data used for automated and accelerated underwriting, or AI and ML models "should not be based in any way on race, color, creed, national origin, status as a victim of domestic violence, past lawful travel, or sexual orientation in any form or manner" (Regalbuto, 2019). This New York circular also places due emphasis on the matter of AI and ML transparency, stating that an applicant is entitled to transparency on how a carrier arrived at a life insurance coverage decision, including all pertinent sources of information used to arrive at the determination. ***The New York guidance - if applied to other industries - is an illustrative example for companies that are pursuing the "buy" model for how they should work with their technology providers to make their AI models more transparent, and not treat these as a black box.***

Another piece of regulation within the life insurance space that can serve as a template for regulation in other industries originated from the State of Colorado. The Colorado General Assembly adopted bill "SB21-169: Restrict Insurers' Use Of External Consumer Data - Concerning protecting consumers from unfair discrimination in insurance practices" in the assembly's 2021 Regular Session. SB21-169 seeks to prohibit insurers from: "Unfairly discriminating based on an individual's race, color, national or ethnic origin, religion, sex, sexual orientation, disability, gender identity, or gender expression in any insurance practice; or Pursuant to rules adopted by the commissioner of insurance (commissioner), using any external consumer data and information source, algorithm, or predictive model (external data source) with regard to any insurance practice that unfairly discriminates against an individual based on an individual's race, color, national or ethnic origin, religion, sex, sexual orientation, disability, gender identity, or gender expression" (Buckner, Ricks, & Esgar, 2021). The bill goes on to state that "After a stakeholder process, the commissioner shall adopt rules for specific types of insurance, by insurance practice, which rules establish means by which an insurer may demonstrate that it has tested whether its use of an external data source unfairly discriminates based on an individual's race, color, national or ethnic origin, religion, sex, sexual orientation, disability, gender identity, or gender expression. Any such rules shall not become effective until January 1, 2023, at the earliest, for any type of insurance. The rules must require each insurer to:

- Provide information to the commissioner concerning the external data sources used by the insurer in the development and implementation of algorithms and predictive models for a particular type of insurance and insurance practice;
- Provide an explanation of the manner in which the insurer uses external data sources for the particular type of insurance and insurance practice;
- Establish and maintain a risk management framework that is reasonably designed to determine, to the extent practicable, whether the insurer's use of external data sources unfairly

discriminates against individuals based on their race, color, national or ethnic origin, religion, sex, sexual orientation, disability, gender identity, or gender expression;

- Provide an assessment of the results of the risk management framework and actions taken to minimize the risk of unfair discrimination, including ongoing monitoring; and
- Provide an attestation by the insurer's chief risk officer that the insurer has implemented the risk management framework appropriately on a continuous basis.
- The rules adopted by the commissioner must include provisions establishing:
- A reasonable period of time for insurers to remedy any unfairly discriminatory impact in an external data source; and
- The ability of insurers to use external data sources that have been previously assessed by the division of insurance (division) and found not to be unfairly discriminatory" (Buckner, Ricks, & Esgar, 2021).

The Colorado legislation offers the clearest sense of a de facto model that other states could use as an outline, with the Commonwealth of Virginia reportedly also evaluating similar legislation based on guidelines issues by the State of Colorado. The Colorado regulation is focused on protecting consumers from potential bias and proxy discrimination, and it is not a giant leap to imagine what similar legislation might look like in other industries.

While being explicit that the regulation is not suggestive that these types of issues exist, the law expects that companies test the use of external data, and AI models, in order to ensure that consumers are not adversely impacted by the potential of unfairly discriminatory results. The law defines external data as, "A data or an information source that is used by an insurer to supplement traditional underwriting or other insurance practices or to establish lifestyle indicators that are used in insurance practices. External consumer data and information source includes credit scores, social media habits, locations, purchasing habits, home ownership, educational attainment, occupation, licensures, civil judgments, and court records" (Buckner, Ricks, & Esgar, 2021). The regulation states that the intent is to expose any issues and resolve these issues during testing, thereby shielding consumers from inadvertent bias resulting from these issues.

Similarly, the NAIC's Accelerated Underwriting Working Group states that "the use of accelerated underwriting in life insurance should be fair and transparent to regulators, consumers, and policymakers. Companies must operate in compliance with applicable laws, and the process and data companies use need to be secure. To accomplish these objectives, regulators should dialogue with consumers, life insurers, and third-party vendors to determine if consumer data is being used in problematic or unfair ways or generating unfair outcomes" (NAIC, 2022). This committee's draft guidelines recommends: "Insurers and other parties involved in accelerated underwriting in life insurance should:

- Take steps to ensure data inputs are transparent, accurate, reliable, and the data itself does not have any unfair bias.
- Ensure that the use of external data sources, algorithms or predictive models are based on sound actuarial principles with a valid explanation or rationale for any claimed correlation or causal connection.
- Ensure that the predictive models or machine learning algorithm within accelerated underwriting has an intended outcome and that outcome is being achieved.
- Ensure that the predictive models or machine learning algorithm achieve an outcome that is not unfairly discriminatory" (NAIC, 2022).

Regulatory Frameworks to Watch

Given that regulation in this field is still developing, companies will continue to be diligent in their use of AI models since there is yet to be standard guidance across industries on the safe and effective use of AI. Summarized here are two of the most prominent frameworks to watch – the European Union AI Regulation as of Q4 of 2023, and President Joe Biden's Executive Order on Artificial Intelligence that was also issued in Q4 of 2023. Note that one unitary, overarching AI regulation will be inadequate for your industry and company. It is expected that these overarching AI regulations form the basis for industry-specific AI regulations as AI continues to mature across industries.

a. The EU AI Act

Similar to the leadership they demonstrated with the European Union (EU) data and information privacy protection regulation known as General Data Protection Regulation (GDPR), the EU led the charge in formulating the world's first regulatory framework for AI. First drafted in April of 2021, the EU became the first major world body to enact AI regulation, by reaching "a provisional deal on landmark European Union rules governing the use of artificial intelligence including governments' use of AI in biometric surveillance and how to regulate AI systems such as ChatGPT" (Chee, Coulter, & Mukherjee, 2023) on December 8th 2023. The regulation is expected to go into force in 2024, after it has been ratified between EU countries and European Parliament members. The regulation will be applicable two years post going into force.

According to an article from Reuters, the regulatory framework "requires foundation models such as ChatGPT and general purpose AI systems (GPAI) to comply with transparency obligations before they are put on the market. These include drawing up technical documentation, complying with EU copyright law and disseminating detailed summaries about the content used for training. High-impact foundation models with systemic risk will have to conduct model evaluations, assess and mitigate systemic risks, conduct adversarial testing, report to the European Commission on serious incidents, ensure cybersecurity and report on their energy efficiency. GPAIs with systemic risk may rely on codes of practice to comply with the new regulation. Governments can only use real-time biometric surveillance in public spaces in cases of victims of certain crimes, prevention of genuine, present, or foreseeable threats, such as terrorist attacks, and searches for people suspected of the most serious crimes. The agreement bans cognitive behavioral manipulation, the untargeted scrapping of facial images from the internet or CCTV footage, social scoring, and biometric categorization systems to infer political, religious, philosophical beliefs, sexual orientation and race. Consumers would have the right to launch complaints and receive meaningful explanations while fines for violations would range from 7.5 million euros ($8.1 million) or 1.5% of turnover to 35 million euros or 7% of global turnover" (Chee, Coulter, & Mukherjee, 2023).

The EU AI Act provides a legal framework to govern "the sale and use of artificial intelligence in the EU. Its official purpose is to ensure the proper functioning of the EU single market by setting consistent standards for AI systems across EU member states" (Hoffmann, 2023). An article published by the Center for Security and Emerging Technology (CSET) provides a succinct overview of the crux of the EU AI Act: "At the heart of the proposal stands its risk categorization system, whereby AI systems are regulated based on the level of risk they pose to the health, safety and fundamental rights of a person. There are four categories of risk: unacceptable, high, limited and minimal/none. The greatest oversight and regulation envisioned by the AI Act focuses on the unacceptable and high-risk categories" (Hoffmann, 2023).

These four categories of AI risk are depicted in Figure 44, image courtesy of Ernst & Young (EY) Switzerland (Sathe & Ruloff, 2023). These four categories are described in their associated regulatory articles. Unacceptable Risk, the highest level of risk, is defined in Article 5, High Risk in Article 6, Limited Risk in Article 52, and Minimal Risk is described in Article 69 of the AI Act.

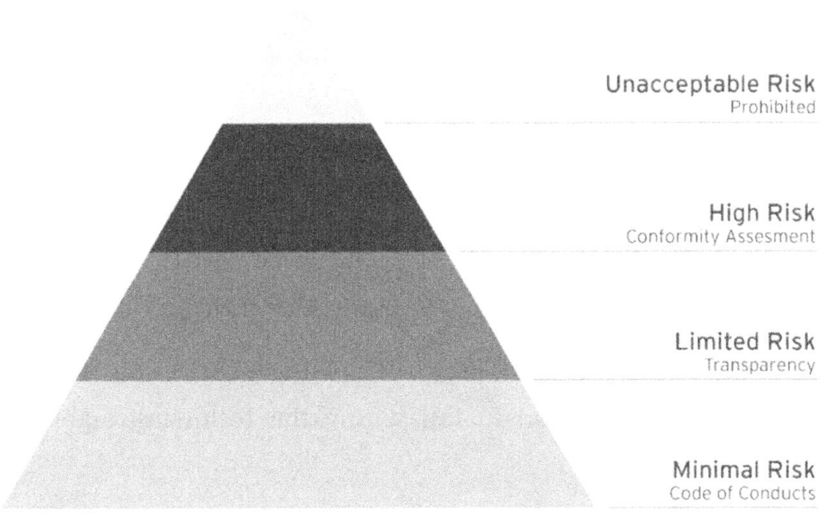

Figure 44: Risk Categories according to the EU AI Act (image credit EY Switzerland)

An article by CSET Researcher Mia Hoffmann states that "Unacceptable Risk Systems will be Prohibited - AI systems belonging to the unacceptable risk category are prohibited outright. Based on consensus between the three proposals, unacceptable risk systems include those that have a significant potential for manipulation either through subconscious messaging and stimuli, or by exploiting vulnerabilities like socioeconomic status, disability, or age. AI systems for social scoring, a term that describes the evaluation and treatment of people based on their social behavior, are also banned. The European Parliament further intends to prohibit real-time remote biometric identification in public spaces, like live facial recognition systems, alongside other biometrics and law enforcement use cases" (Hoffmann, 2023). Stating that high risk systems will be subject to careful regulation, the article categorizes that high-risk AI systems as:

1. "System is a safety component or a product subject to existing safety standards and assessments, such as toys or medical devices; or,
2. System is used for a specific sensitive purpose. The exact list of these use cases is subject to change during the negotiations, but are understood to fall within the following eight high-level areas:
   - Biometrics
   - Critical infrastructure

- Education and vocational training
- Employment, workers management and access to self-employment
- Access to essential services
- Law enforcement
- Migration, asylum and border control management
- Administration of justice and democratic processes" (Hoffmann, 2023).

Organizational Readiness

In a contrast with President Biden's Executive Order on AI, the EU AI Act imposes penalties for noncompliance. According to an article from EY Switzerland, "The penalties for non-compliance with the AI Act are significant and can have a severe impact on the provider's or deployer's business. They range from €10 million to €40 million or 2% to 7% of the global annual turnover, depending on the severity of the infringement" (Sathe & Ruloff, 2023). This same article provides an excellent guide to organizations on how they can prepare their companies for complying with the EU AI Act. This three-step guide – with credit to EY Switzerland – is as below. Refer to Figure 44 for risk categorization.

"Step 1: Model inventory – understanding the current state

To understand the implications of the EU AI Act, companies should first assess if they have AI models in use and in development or are about to procure such models from third-party providers and list the identified AI models in a model repository. Many financial services organizations can utilize existing model repositories and the surrounding model governance and add AI as an additional topic.

Organizations which have not needed a model repository so far should start with a status quo assessment to understand their (potential) exposure. Even if AI is not used at present, it is very likely that this will change in the coming years. An initial identification can start from an existing software catalogue or, if this is not available, with surveys sent to the various business units.

Step 2: Risk classification of models

Based on the model repository, the AI models can be classified by risk. The EU AI Act distinguishes different risk categories:

The Act lays out examples of models posing an unacceptable risk. Models falling into this category are prohibited. Examples include the use of real-time remote biometric identification in public spaces or social scoring systems, as well as the use of subliminal influencing techniques which exploit vulnerabilities of specific groups.

High-risk models are permitted but must comply with multiple requirements and undergo a conformity assessment. This assessment needs to be completed before the model is released on the market. Those models are also required to be registered in an EU database which shall be set up. Operating high-risk AI models requires an appropriate risk management system, logging capabilities and human oversight respectively ownership. There shall be proper data governance applied to the data used for training, testing and validation as well as controls assuring the cyber security, robustness and fairness of the model. Examples of high-risk systems are models related to the operation of critical infrastructure, systems used in hiring processes or employee ratings, credit scoring systems, automated insurance claims processing or setting of risk premiums for customers.

The remaining models are considered limited or minimal risk. For those, transparency is required, i.e., a user must be informed that what they are interacting with is generated by AI. Examples include chat bots or deep fakes which are not considered high risk but for which it is mandatory that users know about AI being behind it.

For all operators of AI models, the implementation of a Code of Conduct around ethical AI is recommended.

Step 3: Prepare and get ready

If you are a provider, user, importer, distributor or affected person of AI systems, you need to ensure that your AI practices are in line with these new regulations. To start the process of fully complying with the AI Act, you should initiate the following steps: (1) assess the risks associated with your AI systems, (2) raise awareness, (3) design ethical systems, (4) assign responsibility, (5) stay up-to-date, and (6) establish a formal governance. By taking proactive steps now, you can avoid potential significant sanctions for your organization upon the Act coming into force" (Sathe & Ruloff, 2023).

Special Note: Companies should be aware that the EU AI Act presents a unique challenge for global and multinational organizations seeking to be globally compliant with the letter, as well as the spirit, of the law. There are certain AI implementations that are prohibited under the EU AI Act that are quite commonplace and considered acceptable in some other countries. An example of this would be AI for facial recognition and social scoring. These uses of AI are prohibited according to the EU AI Act (Unacceptable Risk), but are actually commonplace, and part of the social fabric of other countries like China.

b. President Biden's Executive Order on AI

President Biden issued an Executive Order (EO) on AI on October 30th, 2023. The goal of the Executive Order, as the administration described, is to promote "safe, secure, and trustworthy development and use of artificial intelligence" (The

Biden Administration, 2023). According to a White House Fact Sheet regarding this Executive Order, "The Executive Order establishes new standards for AI safety and security, protects Americans' privacy, advances equity and civil rights, stands up for consumers and workers, promotes innovation and competition, advances American leadership around the world, and more" (The Biden Administration, 2023).

An EY article published the day after this EO was issued, provides commentary on the tenets of the EO as follows: "This Executive Order represents a significant contribution to the subject of accountability in how AI is developed and deployed across organizations. Given the breadth of recommendations and actions provided, it is likely to have an effect on organizations across all sectors of the economy, from the most mature AI implementers to first-time adopters. The Executive Order's definition of AI systems is also broad; it is not limited to generative AI or systems leveraging neural networks but is inclusive of systems which have been built over the last several years.

Determining the extent to which the EO affects an organization will involve careful assessment of not only an entity's own use of AI, but also the extent to which its products and services incorporate or are reliant on third-party vendors' AI-enabled capabilities.

Importantly, the National Institute of Standards and Technology (NIST) will be foundational in the development of guidelines and best practices for "developing and deploying safe, secure and trustworthy AI systems," and companies may consider evaluating their existing AI risk management frameworks against the NIST AI Risk Management Framework to develop a baseline and prepare for additional guidance to be released from relevant agencies and regulatory bodies" (Neill, Hallmark, Jackson, & Diasio, 2023). This article also provides a summary of the Key Takeaways of President Biden's EO on AI, which is guided by eight principles and priorities. This summary, credit EY (Neill, Hallmark, Jackson, & Diasio, 2023), is presented in Figure 45.

1	AI must be safe and secure by requiring robust, reliable, repeatable and standardized evaluations of AI systems, as well as policies, institutions, and, as appropriate, mechanisms to test, understand, and mitigate risks from these systems before they are put to use.
2	The US should promote responsible innovation, competition and collaboration via investments in education, training, R&D and capacity while addressing intellectual property rights questions and stopping unlawful collusion and monopoly over key assets and technologies.
3	The responsible development and use of AI require a commitment to supporting American workers though education and job training and understanding the impact of AI on the labor force and workers' rights.
4	AI policies must be consistent with the advancement of equity and civil rights.
5	The interests of Americans who increasingly use, interact with, or purchase AI and AI-enabled products in their daily lives must be protected.
6	Americans' privacy and civil liberties must be protected by ensuring that the collection, use and retention of data is lawful, secure and promotes privacy.
7	It is important to manage the risks from the federal government's own use of AI and increase its internal capacity to regulate, govern and support responsible use of AI to deliver better results for Americans.
8	The federal government should lead the way to global societal, economic and technological progress including by engaging with international partners to develop a framework to manage AI risks, unlock AI's potential for good and promote a common approach to shared challenges.

Figure 45: Key Takeaways from President Biden's Executive Order on AI (October 30th, 2023) – credit EY

Anjana Susarla, professor of information systems at Michigan State University, provided an analysis of this EO on PBS' "The

Conversation": "Researchers of AI ethics have long cautioned that stronger auditing of AI systems is needed to avoid giving the appearance of scrutiny without genuine accountability. As it stands, a recent study looking at public disclosures from companies found that claims of AI ethics practices outpace actual AI ethics initiatives. The executive order could help by specifying avenues for enforcing accountability.

Another important initiative outlined in the executive order is probing for vulnerabilities of very large-scale general-purpose AI models trained on massive amounts of data, such as the models that power OpenAI's ChatGPT or DALL-E. The order requires companies that build large AI systems with the potential to affect national security, public health or the economy to perform red teaming and report the results to the government. Red teaming is using manual or automated methods to attempt to force an AI model to produce harmful output – for example, make offensive or dangerous statements like advice on how to sell drugs. Reporting to the government is important given that a recent study found most of the companies that make these large-scale AI systems lacking when it comes to transparency.

Similarly, the public is at risk of being fooled by AI-generated content. To address this, the executive order directs the Department of Commerce to develop guidance for labeling AI-generated content.

Federal agencies will be required to use AI watermarking – technology that marks content as AI-generated to reduce fraud and misinformation – though it's not required for the private sector. The executive order also recognizes that AI systems can pose unacceptable risks of harm to civil and human rights and the well-being of individuals: "Artificial Intelligence systems deployed irresponsibly have reproduced and intensified existing inequities, caused new types of harmful discrimination, and exacerbated online and physical harms."

What the executive order doesn't do

A key challenge for AI regulation is the absence of comprehensive federal data protection and privacy legislation. The executive order only calls on Congress to adopt privacy legislation, but it does not provide a legislative framework. It remains to be seen how the courts will interpret the executive order's directives in light of existing consumer privacy and data rights statutes. Without strong data privacy laws in the U.S. as other countries have, the executive order could have minimal effect on getting AI companies to boost data privacy. In general, it's difficult to measure the impact that decision-making AI systems have on data privacy and freedoms.

It's also worth noting that algorithmic transparency is not a panacea. For example, the European Union's General Data Protection Regulation legislation mandates "meaningful information about the logic involved" in automated decisions. This suggests a right to an explanation of the criteria that algorithms use in their decision-making. The mandate treats the process of algorithmic decision-making as something akin to a recipe book, meaning it assumes that if people understand how algorithmic decision-making works, they can understand how the system affects them. But knowing how an AI system works doesn't necessarily tell you why it made a particular decision" (Susarla, 2023).

Chapter Twenty-Three: Ten Enterprise-Level Best Practices – Part 7

This chapter presents an overview of Enterprise Best Practices 6 through 10. While these are critically important to the success of your enterprise AI program, it is important to prioritize Enterprise Best Practices 1 through 5 before focusing on the five best practices presented here.

Enterprise Best Practice 6: Careful Vetting of External Data

This best practice recommendation applies primarily to companies that have adopted, or are considering adopting, a "build" model for their enterprise AI program.

These companies partner with trusted technology providers to ingest external, non-traditional sources of data for their AI models. These data sources are selected based on the enterprise AI use case in question – for example, your loan processing AI model will ingest potentially hundreds of external data sources, including zip code data from the US Postal Service via an API, as well as median income levels for a particular zip code. Additionally, your algorithm will likely ingest data from external data sources that is specific to the applicant – from commonplace datasets such as credit source, to any other pertinent data sources that your loan processing rules deem to be fit-for-use, and appropriate to issuing a loan to an applicant.

Companies - “early adopters” and “laggards” alike - partner with technology providers to ingest external data sources. Sometimes this selection is informed by the guidance of the technology providers themselves, consultants, or other third-parties, in addition to internal experts and stakeholders. There is careful selection and vetting of the external data source to ensure that the data meets the company’s needs and AI strategy. However, given the inherent trust and established relationship between the company and the data provider, there is often little to no quality checks performed on these data sources during the time of ingestion, as these are considered trusted providers and trusted sources of data.

As evidenced by the best practice on emerging regulation, it is incumbent upon a firm to not just have absolute faith in their technology provider, but also to be able to perform a level of testing on these data sources during ingestion, and be able to furnish the best results to consumers and regulators if asked. In addition to developing a plan around which data sources to consume commensurate to their business value, companies should invest in a data strategy around the consumption of these external data sources, with appropriate quality assurance and quality control rigor.

Enterprise Best Practice 7: Focus on Data Quality and Data Literacy

According to Dr. Sebastian-Coleman, "Data literacy is the ability to read, understand, interpret, and learn from data in different contexts. It also includes the individual's ability to apply what is learned to different contexts, including communicating about data to other people" (Sebastian-Coleman, 2022).

Data is at the crux of AI, and will continue being at the core of any AI-related use cases across your value chain in the future. For an enterprise to enable its AI program for success will require the organization to digitize by traversing across the value chain. This cross-functional digitization implies that each employee in every department that this applicant-centric value chain traversal touches will interact with data at varying levels. From the data scientists that develop a homegrown AI model, or ingest external data sources, to the business unit professionals involved in helping test the efficacy of these models, to those responsible for testing new data sources for potential predictive value, there will be no person that does not consume, create, update, or view a wide variety of pertinent data.

With the focus on data quality, none of these individuals would be able to discern good data from bad data without being data literate. Without being data literate, employees across the value chain will be unable to interact with data in any meaningful manner, nor would they be able to endow this data with context and meaning. With firms prioritizing ethical issues over methodological issues, data literacy is foundational to equip employees to make ethical AI decisions. Data literacy is vital to data strategy and governance and to shift the organization's mindset towards being a data-driven culture.

Enterprise Best Practice 8: Working with Technology Providers

Companies - regardless of if they leverage a "build" or "buy" model – typically enjoy a strong relationship with their technology providers. Technology providers across industry ecosystems are established players, with robust data governance and mature data management practices. They inspire implicit trust in their customers (partner companies), and are generally considered trusted data sources.

When firms employ the "build" model, there is often limited vetting of the quality of data being received from these technology providers. Technology providers that offer a full-service suite - that is, they ingest a company's AI-use case pertinent data and business rules, and provide back to the firm an AI output, such as a prediction, or recommendation – are generally treated by companies as an AI "black box." Going forward, especially considering emerging regulations, it will be vital for companies to work with, and truly *partner* with, their technology providers to clearly understand sources of data being leveraged to reach an AI decision.

Your company should expect to be held accountable to explain – if/when asked by a consumer or regulator - how a particular decision was reached. In this instance, firms cannot simply point to their technology providers to furnish this explanation. Companies often do not have a line of sight into data sources being leveraged by technology providers, to arrive at a decision. It is highly unlikely that a technology provider would divulge details regarding their proprietary AI models to a company even when asked.

At the least, it will be important for carriers to understand what sources of data are contributing into arriving at an AI decision, how the technology provider ensures data quality across these data sources, and what steps this technology provider undertakes to eliminate the chance of potential bias and proxy discrimination from seeping into their predictive models.

Enterprise Best Practice 9: Clearly Defined Roles and Responsibilities across Enterprise

Clearly defined roles and responsibilities are the linchpin to equip any enterprise related data programs for success. Roles and responsibilities are a bedrock of data strategy and governance programs, including establishing the concept of data stewards across the organization. When it comes to an enterprise AI program, and in taking a customer-centric view across your enterprise value chain, it will be vital for your firm to clearly identify who is responsible for what, when, and what specific actions and responsibilities they have in the value chain.

There are no clear models in place across industries to define roles and responsibilities as pertains to AI within an enterprise as of the first quarter of the 21st century. These responsibilities are distributed across multiple teams, from individual business units, to IT, and data science. The answers are unclear even when it comes to the topic of who is ultimately responsible and accountable for an organization's business-unit level AI initiatives. As part of developing a holistic strategy and roadmap for your enterprise AI program (including divisional/departmental AI programs), it is strongly encouraged that an organization invest the time in developing a RACI matrix commensurate to their AI program. The RACI model, which stands for "Responsible," "Accountable," "Consulted," and "Informed" adds structure and role clarity that individuals play within a project by clarifying responsibilities and ensuring everything that needs to be accomplished within a program has an individual assigned to it.

An article in CIO Magazine further clarifies the four roles that individuals might play in any program within context of the RACI

Model as "i. *Responsible*: People or stakeholders who do the work. They must complete the task or objective or make the decision. Several people can be jointly responsible. ii. *Accountable*: Person or stakeholder who is the "owner" of the work. He or she must sign off or approve when the task, objective or decision is complete. This person must make sure that responsibilities are assigned in the matrix for all related activities. Success requires that there is only one person Accountable, which means that "the buck stops there." iii. *Consulted*: People or stakeholders who need to give input before the work can be done and signed-off on. These people are "in the loop" and active participants. iv. *Informed*: People or stakeholders who need to be kept "in the picture." They need updates on progress or decisions, but they do not need to be formally consulted, nor do they contribute directly to the task or decision" (Kantor, 2022). The RACI chart will help in identifying roles and responsibilities across the organization, and is an effective communication tool to ensure that these resources are aware of their roles and specific responsibilities in the enterprise AI value chain.

Enterprise Best Practice 10: Structural Setup for Sustained Success

Just as roles and responsibilities across an organization when it comes to an enterprise AI program are maturing across industries, there is no clear or right answer on how to structurally set up an organization in the best manner for AI success.

Drawing from the successful models being employed for data strategy and governance, the AIM Framework© recommends either establishment of a structure similar to those used for data strategy and governance, which might be appropriate for more mature organizations. Firms should at the least consider augmenting their existing data strategy and governance programs to encompass responsibilities related to AI. Some organizations are investing in expansion of data governance programs to include AI and ML, sometimes clumped under the "Analytics" function. It is worth considering that organizations with a formal "Head of Data" or "Head of Analytics" role, such as a Chief Data Officer (CDO), or a Chief Data Analytics Officer (CDAO), have their responsibilities be expanded to encompass AI and ML. Organizations that achieve a maturity model of Level 04 and beyond might be well served to consider the creation of a Chief Artificial Intelligence Officer (Chief AI Officer) role, a position that could reside side-by-side with the Chief Data Officer, or one that reports into the CDO with dotted line responsibilities into the line-of business heads (C-suite) as applicable, and your Chief Information Officer.

For companies that operate in a matrixed environment, their AI programs are staffed and resourced accordingly - a cross-functional representation of individuals on a scrum team that report up through varied divisions. While this model has been effective and works fine to launch and stabilize programs, it is unlikely one that will allow organizations to achieve scale. Unless an organization decides to spin up multiple scrum teams in a Scaled Agile Framework (SAFe) model, it will be important to consider, especially for organizations that pursue the "build" model, to have dedicated teams that are fully aligned to the enterprise AI program.

Another frequent challenge that companies face is that resources on scrum teams are not full-time on AI-related programs. This is especially true in the case of shared teams, where resources are susceptible to being reallocated to other projects across the company. Analytics teams are especially vulnerable to this, given that their services continue to be in high demand throughout a company. The crux of enterprise AI programs, especially with the "build" model, is not just about development and launching of the program, but it's the ongoing care, feeding, and maintenance of these machine learning models. It is therefore vital to have a scrum team that is fully committed and dedicated to your prioritized set of AI use cases. Larger companies should consider regular "scrum of scrums" or a formal mechanism that allows their data science, AI and ML teams to socialize and share information with each other. There is limited knowledge sharing that happens across multiple lines of business in larger firms, despite having similar opportunities and challenges with all things AI. Having a central executive, such as a Chief Data Officer with expanded responsibilities, or a Chief AI Officer, would alleviate some of the challenges that come with organizations not freely sharing information and knowledge within the same company.

With the ten Enterprise Best Practices serving as a foundation, the next chapter will outline five best practice recommendations around each: the People, Process, and Technology dimensions of the PPT Framework.

Chapter Twenty-Four: People/Process/Technology Best Practice Recommendations – Part 1

With the ten Enterprise Best Practices as a predicate, this chapter delves into best practice recommendations around the People, Process, and Technology dimensions of the PPT Framework.

A tabular depiction of these fifteen PPT best practices, five within each domain, are depicted in Figure 46 below. The **AI and ML Governance Model** is highlighted as a key finding in the grid below:

PEOPLE	PROCESS	TECHNOLOGY
1. Industry Domain Knowledge across Value Chain	1. Processes that Promote Transparency and Explainability	1. Standard IT Supported Technology Stack
2. Regulatory Knowledge across Value Chain	2. Customer Education on Use of Personal Data	2. 3rd Party Data Provider Taxonomy
3. Dedicated Teams and Resources	3. Documented, Repeatable Data Selection Processes	3. Automated Testing for Inadvertent Bias and Proxy Discrimination
4. AI and ML Governance Model ★	4. Documented, Repeatable Technology Provider Selection Processes	4. Automated QC Processes
5. Knowledge Sharing across Enterprise	5. Testing Rigor, Proactive Publication of Audit Results	5. APIs and Cloud as Core Requirements

Figure 46: Overview of People/Process/Technology Best Practices

People Dimension Best Practices

Effectuating AI best practices into the very core of your enterprise as a strategic enabler is less about technology, and more about your organizational culture. Your people are your most important strategic asset to help enable this cultural transformation. You can only be successful at your AI strategy with an undivided focus on your people. Figure 47 below outlines the criticality of the People dimension in your AI strategy. Note that although the AIM Framework© outlines five best practices for the People dimension, you are encouraged to augment these with your own corporate best practices regarding culture and change management that are centered around your employees.

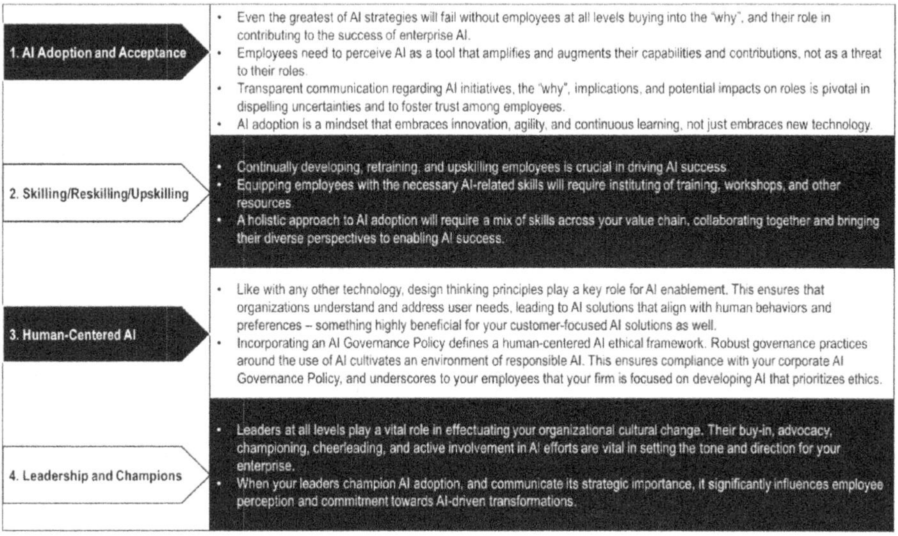

Figure 47: The Vital Need to Focus on People

People Best Practice 1 - Industry Domain Knowledge across Value Chain

Several mature and established industries struggle with attracting and retaining talent. With an aging existing workforce, companies seek to attract employees from different industries and sectors into their industry. Not having been exposed to the new industries that they are entering, these new employees enter these organizations with fresh ideas and innovative thinking, but lack the industry technical knowledge to put these innovative ideas into practice. This causes these individuals to drop out of that industry and leads to retention challenges, which in turn results in companies seeking to bring in talent from outside the industry, in an infinitely repeating closed loop.

The post-COVID economic recovery was marked by historically low unemployment rates, especially in the United States. This led to what has been termed as the "Great Resignation," and resulted in many non-digitally native companies having been severely challenged in being able to fill all their open positions. This issue has been especially acute in skilled trades such as IT, and Data Science. Exacerbating the issue is that adoption of AI is rapidly expanding across all industries – and IT professionals continue to have a choice, presenting them with unlimited choices beyond any one industry. A point of consideration for organizations considering "build" versus "buy" is to keep in mind that there continue to be a limited number of AI subject matter experts and practitioners within every industry. The talent pipeline is severely constricted and there is a lack of talent to develop new and maintain existing AI models.

There is a direct correlation between investing in employees' education, training and development, and talent attraction and retention (Chen, 2014). Your general attitude towards investment in employee development should be that you should strive to make an individual's resume look better when they leave your organization than when they first came to your organization. The AIM Framework© recommends that companies invest significantly in not just onboarding new talent into their industry, but truly in investing in their learning of the industry. This is not just a talent retention strategy, but will also significantly help in extending the work of your AI strategy. New talent with fresh and innovative thinking needs to be grounded with industry knowledge to be able to truly develop innovative new ideas that can be operationalized and workable inside an organization.

Companies would be well-served by engaging with their industry trade associations, and investing in industry education to inspire a broader picture perspective of how the industry operates to employees unfamiliar with the industry. When AI teams, such as data scientists, are focused on the immediate use cases they are working on, without broader knowledge of the industry, they might be missing out on other opportunities and extensions of AL and ML across the company. This "penicillin moment" (Gaynes, 2017) will continue to evade companies and stifle innovation without sustained and concerted efforts to educate employees across the value chain on the basics of your industry.

People Best Practice 2 - Regulatory Knowledge across Value Chain

Regardless of if your company operates within a highly regulated industry, your organization is likely subject to some level of regulatory and compliance requirements. Whether at the state, federal, or international level, these regulations are poised to be much broader and deeper when it comes to enabling your business with AI. There might be facets of your business today that are seemingly innocuous and not subject to any special regulatory requirements. However, the same lines of business that are innocuous today, might be considered from an AI compliance aspect in the future. It is vital that associates across your value chain that are working on your enterprise AI strategy, have a clear understanding of these regulatory implications, especially when it comes to your industry, and specifically when it comes to the use of AI within your industry.

In addition to investing in basic industry knowledge, companies would be well served in teaching their employees across the value chain about regulatory and compliance issues. Not only do these teams need to be well versed in understanding the fairness and ethical concerns around use of AI, but they also need to be savvy enough to recognize what they can and cannot do within the bounds of existing regulatory guidance.

Another area that companies need to pay attention to is around data privacy and protection. While there are no Federal guidelines around data privacy, firms, especially multinationals and those that operate in multiple states, need to pay special attention to data privacy regulations such as GDPR in Europe and the UK, PIPL in China, CCPA and CPRA in California, and other state-based data privacy regulations. The central predicate of these data privacy regulations are around a consumer's right to exercise control over what data can be shared about them, and with whom. While this is a regulatory concern that is not limited to one facet of your business, the classifications of data that your organization might collect on your customers – Sensitive PII (SPII), NPPI, HIPAA, etc. – implies a level of compliance complexity that warrants all employees understand the regulations and what it means to your company.

Data privacy regulations include the consumer's right to opt in or opt out of a company's ability to share their data, their right to have their information corrected, their right to have their information purged, and the right to be able to port their data. Between regulations around data privacy and protection, and other regulatory proposals on AI use, it is vital for every employee across the cross functional value chain of an organization, to be familiar with regulatory and compliance constraints that their work is subject to.

This is also an area where companies would be well served to invest in regulatory and compliance training for their employees from their industry trade associations. Lastly, in addition to training on the regulatory and compliance landscape, the AIM Framework© recommends that organizations continue to invest and expand ethics training across the employee base. This training should expose and align employees to a common taxonomy of inadvertent bias and proxy discrimination within the context of AI models and how each employee can be a safeguard in detection, identification, and raising concerns.

People Best Practice 3 - Dedicated Teams and Resources

Across industries, cross-functional teams such as those within agile scrum teams, are susceptible to context-switching and being moved off an AI project to work on other enterprise priorities. This occurs regardless of if the company is considered to be an "early adopter," or a "laggard," or pursued a "build" versus "buy" model for AI. This has specifically been highlighted as an issue within data science, data analytics, and IT teams.

Companies that leverage a Center of Excellence (CoE) model for their data analytics teams, where there is a pool of shared resources who are "farmed out" to various agile scrum teams across the organization, are especially susceptible to being pulled from one project and reassigned to another as priorities dictate. This context switching is detrimental to creating a depth of subject matter expertise in any domain-specific prioritized AI initiatives. For large companies with multiple lines of business, this could mean that the pool of data scientists can never truly be subject matter experts and practitioners in any one domain.

The lack of dedicated teams also results in projects being potentially beset with unexpected delays. It can take approximately 3 to 6 months for updates to existing data sources to be consumed within a company's AI model (for one model), and an average of a full year to onboard a new data source. Even for firms that pursue a "buy" model, it can take an average of 3 to 6 months to understand what updates their technology provider partners are making to the AI models, and whether these updates comport with their business need. In such scenarios, it is important to retain a dedicated team. As mentioned, AI models, especially those built by companies ("build" approach), require continual care and maintenance. Model drift can occur without having a team of experts continue to monitor and refine the models. Without having dedicated resources allocated to enterprise AI programs, firms continue to expose themselves to the risk of their models going into disrepair, or for their programs not delivering the business value they seek.

The AIM Framework© strongly recommends that companies consider having dedicated teams assigned to their AI programs, during the resourcing phase, as a part of developing their plans, strategy, and roadmap.

People Best Practice 4 - AI and ML Governance Model {⚡ KEY FINDING/RECOMMENDATION}

There continues to be broad disparity across many industries, especially established and mature ones, on the number of companies within an industry with active data strategy and governance programs.

Some of this disparity can be attributable to the fact that organizations have a spectrum of definitions of what data strategy and governance means – and more importantly, what it means to them. For some, it is as simple as a central repository such as an enterprise data warehouse and mastery of the data that this repository contains. For others, it is the full gamut of organizational mobilization that is at the heart of sustained success of these programs in the quest to transform an enterprise to being a data-driven one. According to Randy Bean who cites barriers for organizations to become data-driven, "First, achieving data-driven leadership remains an aspiration for most organizations - just 26.5% of organizations report having established a data-driven organization. Second, becoming data-driven requires an organizational focus on cultural change. In this year's survey, 91.9% of executives cite cultural obstacles as the greatest barrier to becoming data-driven. As noted, this is not a technology issue. It is a people challenge. Lastly, organizations are establishing the leadership function - in the role of the Chief Data and Analytics Officer - which will provide the foundation for becoming data-driven. However, just 40.2% of companies report that the role is successful and well established within their organization" (Bean, 2022).

While some studies cite that "64% of companies have data governance programs" (Anandarajan & Jones, 2021), others put this figure at around 50% (erwin & DATAVERSITY, 2020). A majority of large to midsized organizations have invested in data strategy, governance, and management programs – with the larger and better-funded companies having adopted a formal governance model. Most of these companies also report a myriad of challenges with their programs that serve as barriers from being able to realize the programs' full potential. These challenges include securing sustained C-suite and CEO support and commitment, influencing business process changes across the enterprise, prevailing cultural challenges, change management, etc.

As the challenges with data governance and management indicate, instituting a governance model for an ongoing and mature process is a lot more challenging than it would be for implementing a model at the start, or relatively soon after commencement of a practice. While organizations continue to manage and overcome their data governance and management challenges, the AIM Framework© recommends that an organization's AI practice be equipped with an AI Governance and AI Management framework. The AIM Framework© defines AI Management as decisions that an organization makes around AI, with AI Governance providing the structural support that allows an organization to make these decisions.

As referenced in the Enterprise-Level Best Practices, the AI Governance recommendations presented here are intended to propose what an organizational structure of such a framework might look like. This framework is intentionally generic to allow firms to co-opt the recommendations and customize it to suit their organizational needs. Figure 48 depicts the three primary organizational components of an AI Governance Triad. The central

AI Governance Office enjoys active C-suite involvement and support, and is comprised of members of the C-suite (if not the CEO) and the Chief Data Officer or the Chief AI Officer.

The AI Governance Council is chaired by the Chief Data Officer/Chief AI Officer, and is composed of line of business leaders across the automated and accelerated underwriting value chain. These would be individuals who are one level below their respective C-suite divisional heads, such as a leader from the marketing department who reports to the Chief Marketing Officer, the individual reporting to the Chief Financial Officer, etc.

The AI Domain Stewards are representatives from each division in the value chain who bear responsibility for their part in the data and/or AI models. When it comes to responsibility over data used for the prioritized AI use case/s, these could be the Data Stewards themselves, bearing responsibility to ensure that data being ingested into AI models is subject to appropriate data governance practices as established by the AI Governance Office. As far as AI Domain Stewards for the AI/ML models themselves, they could be a distinct individual from the Data Steward/s for that department, or depending on size and maturity of the firm, could be a responsibility that is collapsed into the same individual/s.

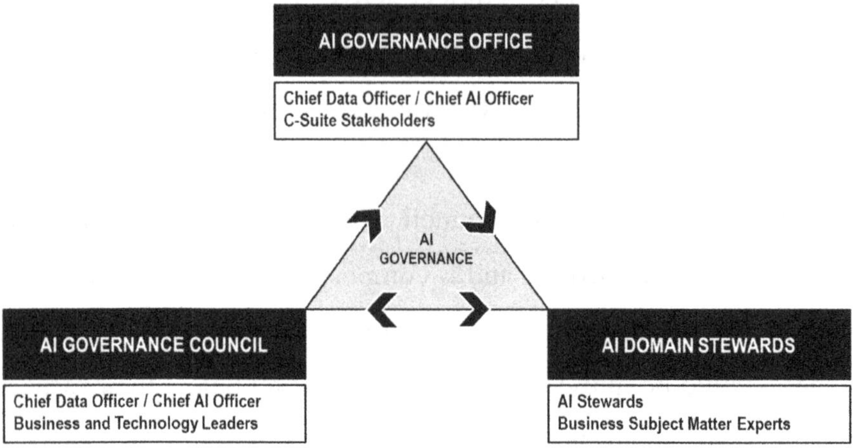

Figure 48: Primary Organizational Components of the AI Governance Triad

A matrix of expected responsibilities of these groups of individuals, by organizational component, within the context of a corporate hierarchy is depicted in Figure 49. It also outlines high-level expected responsibilities for each of these AI Governance domains.

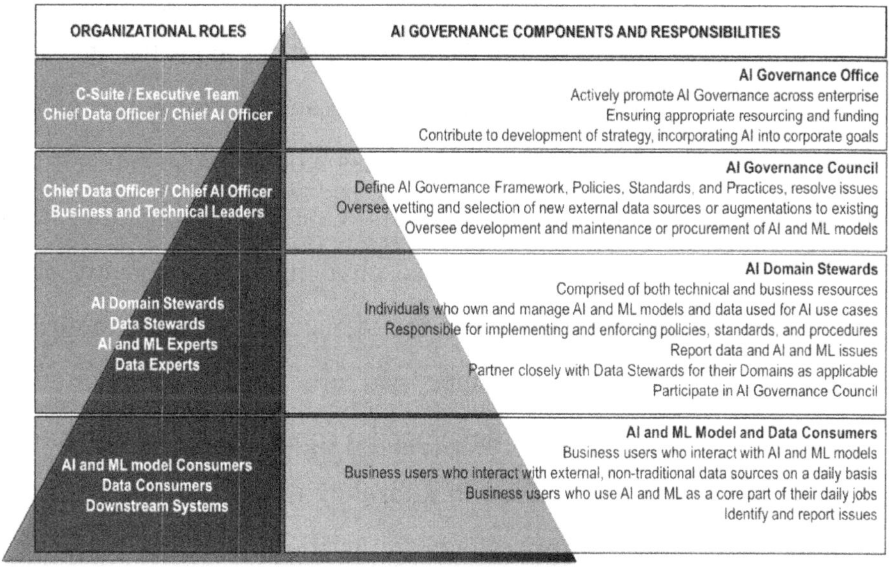

Figure 49: AI Governance Organizational Framework Overview

People Best Practice 5 - Knowledge Sharing across Enterprise

We have covered the fact that most companies across established industries have traditionally been operating in digital silos. Even within a specific line of business, opportunities to share and socialize best practices have historically been limited. These limitations are often self-imposed, exacerbated by disparate systems, distinct processes, and a general reticence to change when dealing across cross-functional departments.

A fundamental facet of enabling an enterprise AI strategy is to effectively traverse horizontally across your organizational value chain. It will therefore be important to share information and best practices within a line of business, and just as important to share across lines of business. Within large organizations, with multiple lines of business, one line of business could greatly benefit in learning from another line of business. One line of business in an organization might have experienced challenges and growing pains that could inform strategies for the departmental AI strategy of another line of business. Groups that are farther ahead in their implementations might have developed a robust set of best practices that can be effectively applied in a turnkey manner within the same organization due to enterprise-level similarities.

Absent an intentional approach to facilitate knowledge sharing across an organization, these opportunities might go unnoticed, underleveraged, and undercapitalized. Taken in aggregate, not being able to freely socialize information across a company can be detrimental to the company's operational efficiencies, and mitigate achievement of scope and scale economies. The AIM Framework© recommends that the Chief Data Officer, or Chief AI Officer, or in the absence of either of these roles, the Chief Information Officer of an organization actively ensure that best practices are shared across the enterprise by cultivating an environment and enabling supporting processes to be able to successfully do so.

A People-Centered AI Strategy

The human factor is at the heart of AI adoption, since success with an AI strategy is fundamentally based on nurturing a corporate culture that thrives amidst change, is adaptable and agile to embrace continual change, and is not afraid to make mistakes and learn from them. These corporate cultures are ones that promote continual learning, a spirit of collaboration, and foster an environment that is conducive to innovation. Their technology ethics are reflective of their corporate credo around ethics, with empowered employees who are engaged in the industry that they serve.

The next chapter shall delve into five Process, and Technology Dimension Best Practices.

Chapter Twenty-Five: People/Process/Technology Best Practice Recommendations – Part 2

The preceding chapter explored five People Dimension Best Practices. This chapter delves into best practice recommendations around the Process and Technology dimensions of the PPT Framework.

Process Dimension Best Practices

Well-developed and well-defined organizational and operational processes are vital to ensure the smooth planning and implementation of your enterprise AI strategy. Well-defined processes ensure organizational and operational standardization and consistency. They allow for optimization, risk management and mitigation, resource efficiencies, and fertilize the enterprise for a continual learning, continual improvement mindset. Organizations that focus on the establishment and fine-tuning of processes for AI to be lucidly integrated into their enterprises lay the groundwork for strategic and sustained AI success. Figure 50 below outlines the importance of well-defined processes that govern AI across companies, serving like a skeleton that supports the body of effective AI implementation, management, and sustained growth.

1. Standardization	• Processes establish standards and guidelines for AI, fostering consistency across various projects and initiatives, across your enterprise value chain. • Standardized processes ensure consistency in data management, model development and training, testing and validation, deployment, and maintenance of models, enhancing predictability and reliability in AI outcomes.
2. Resource Allocation	• Processes facilitate enterprise planning and enable optimal resource allocation and utilization for AI programs, by guiding decisions related to talent allocation, infrastructure requirements, and budget allocation for prioritized AI use cases. • Processes outline procedures for the expansion of your AI program as your organization's needs evolve, thereby allowing for planned scalability.
3. Operational Efficiencies	• Defined processes streamline the implementation of AI, optimizing efficiency and effectiveness. • Structured workflows ensure easier integrations, reducing inefficiencies and redundancies. • Processes enable seamless collaboration among cross-functional teams in your value chain.
4. Risk and Governance	• Processes play a pivotal role in risk management, risk mitigation, and governance by identifying and addressing potential pitfalls, including biases in data or model performance issues, before they manifest into significant challenges. Processes are instrumental in ensuring compliance with regulatory guidelines, helping the firm to adhere to data privacy laws, ethical AI guidelines, and industry-specific regulations, thereby mitigating legal risks and reputational damage.
5. QA and QC	• Processes are the bedrock for quality assurance and quality control in AI. • Well-defined QA/QC measures include rigorous testing, validation, ongoing monitoring, reporting, and remediation of issues, which allows for ensuring the accuracy, reliability, and ethical implementation of AI.
6. Continual Improvement	• Processes allow you to incorporate feedback loops for continual learning and continual improvement. • This helps drive innovation by fostering a failure-resistant culture of learning, allowing firms to rapidly prototype, iterate, incorporate advancements, and adapt to changing business requirements, and/or technological advancements.

Figure 50: The Importance of Well-Defined Enterprise Processes

Process Best Practice 1 - Processes that Promote Transparency and Explainability

Most organizations tend to handle the sensitive topics of transparency and explainability for AI internally. Some are guided by technology partners/vendors and consultants, but for the most part, companies tend not to lean on other firms for advice.

Being aware of disclosure concerns, firms conduct their own internal reviews and include their legal and compliance teams in the process. They also conduct extensive peer reviews with other data scientists within their own organization. Companies invest significant time and resources into developing and maintaining extensive documentation around their AI practices. The AIM

Framework© recommends that as companies enact their enterprise AI strategies, they should consider enabling processes that promote this level of layered internal review and provides the same level of transparency and oversight across the end-to-end value chain.

Process Best Practice 2 - Customer Education on Use of Personal Data

The internal processes of most companies are not easily understood by consumers. Some of the opacity is appropriate – organizations need not expose their customers to their "sausage making." Companies are complex, and their internal processes are complex. Customers do not necessarily care about how a product is delivered to them, as long as it is delivered to them at a cost they can bear. Some internal processes are opaque since they are considered as an organization's intellectual property. Some opacity might be required for regulatory reasons.

However, simplification of the process and educating consumers on what it is that they are buying and the benefits that they derive from the purchase, is going to be key to transparency around AI, specifically if AI is leveraged as an enabler during any part of the company-customer engagement. As companies seek to establish and evolve their AI programs, it will be critical for companies to educate and clearly explain to prospective customers on exactly what data the company is collecting on them, why they are collecting this data, and ultimately how they will be using this data. As data privacy regulations continue to expand, and organizations need to demonstrate compliance across multiple geographies in terms of data privacy, it is anticipated that companies will continue experimentation with new and novel data sources, including those that they had stayed away from for concerns of bias or proxy discrimination. Unstructured data, such as data originating from wearable devices, IoT, etc., wherein consumers lifestyle habits are recorded and could be sent to a company to allow for behavior-based marketing or product development, should be explained clearly to consumers.

The fact that firms will be seeking data on a customer in order to provide them with a better, more personalized service, and that their data will not be used for marketing purposes, or sold, should be explained. A new generation of consumers that is already used to sharing their personal data such as their location or health data with numerous mobile apps today, would likely be willing to share the same with a company as long as it is clearly explained to them.

Process Best Practice 3 - Documented, Repeatable Data Selection Processes

Most companies that can be categorized as "early adopters," had a limited choice of novel external data sources to choose from for their AI models, when they commenced their AI journeys. These electronic data sources continue to increase in quality and quantity.

Firms who have been late to AI have established precedents to follow, and can afford to follow the trails that others have blazed before them. However, regardless of company or industry, it is unclear how robust and mature any particular company's external data vetting and selection processes are. There is no definitive answer on where data selection and vetting originate within a firm. In the case of some companies, it is with data scientists experimenting with their AI and ML models, while ingesting new data sources that they have found through a process of discovery. For others, it was contingent on guidance from their vendors and technology consultants. For some, it was the deepening of an existing relationship with their technology provider, or how effective the technology providers were with their sales and marketing campaigns.

Companies would be well served to invest in documented, and repeatable data vetting and selection processes. These documented and repeatable processes will help significantly in being able to placate some of the emerging regulatory requirements.

Process Best Practice 4 - Documented, Repeatable Technology Provider Selection Processes

Thematically similar to the data selection process, there is no standard and repeatable process on how firms select technology providers.

Regardless of if a company chooses to pursue the "build" model and simply use a technology provider to ingest external sources of data, or if they pursue a partnership with a full-service technology provider, the choice of selecting a specific provider is not a repeatable process that an organization can follow. Although firms tend to heed the guidance of their external consultants, and/or their technology partners, understanding what makes a specific technology provider stand out versus others, is going to be important going forward.

This is driven by the fact that industry ecosystems continue to grow and shift through a series of mergers, acquisitions, and new technology providers continue to appear on the scene. One technology provider might not meet the business needs articulated in a broader vision for the enterprise, and a company might need to evaluate alternative providers. In this case, having a documented, repeatable process would make this vetting and selection easier.

Process Best Practice 5 - Testing Rigor, Proactive Publication of Audit Results

Companies choose to conduct their own internal audits and internal testing on external data sources, as well as their AI and ML models.

While conducting internal audits is helpful for firms to be able to immediately resolve and remediate issues as they come up, they should consider being proactive and ahead of upcoming regulations. The AIM Framework© recommends that firms develop a way to publish results around their audits and testing, implicitly underscoring their rigor around testing of AI and ML models, while preserving details of their proprietary practices.

Technology Dimension Best Practices

AI is a continuation of society's accelerating technological innovation. Implicitly, AI and technology can be considered to be synonymous, and symbiotic with each other. Technology advancements are required to support AI advancements, and advancements in AI will continue to propel technological innovation. Technology – whether it is the hardware infrastructure that AI models run on, cloud platforms that give AI its scale, or the very software programming interfaces required to author AI model code (algorithms) – is the foundation of AI. Figure 51 outlines the vital nature of this concurrently symbiotic and synonymous relationship.

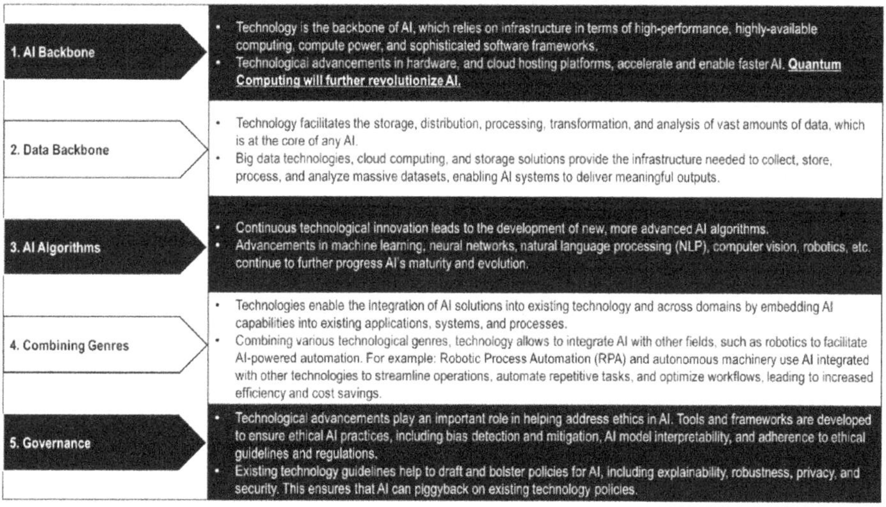

Figure 51: The Synonymous Symbiosis of Technology and AI

Technology Best Practice 1 - Standard IT Supported Technology Stack

Most organizations, regardless of size, could have an onerous number of platforms and technology. Not only do companies struggle with managing technical debt and legacy systems, but most companies' IT organizations also contend with a measurable number of non-IT owned, built, supported, or governed platforms that business users have developed and maintain. These "IT cottage industries" not only pose an operational and potential cybersecurity risk, but they also stand contrary to the cause of standardization and establishment of repeatable processes with predictable outcomes.

Departments such as data analytics, and data science seek to prove the efficacy of their hypotheses and models, by building tools and by experimentation. This experimentation often leads to these groups developing niche products, sometimes using unsupported platforms and technology, and institutionalizing these platforms within their departments. As organizations develop their enterprise AI roadmaps, it will be incumbent upon the Chief Information Officer, the Chief Technology Officer, the Chief Information Security Officer, and the Chief Enterprise Architect of these firms to publish and govern the technology stack that is supported by the firm's IT group.

Technology Best Practice 2 - 3rd Party Data Provider Taxonomy

As mentioned in this book, firms who engage with technology providers in a "buy" model, are often unaware of the type, quantity, and nature of the data sources that a particular technology provider leverages in their AI models in order to render a recommendation, or an expected AI output.

This will prove to be a challenge with emerging regulations on the matter, and firms would serve themselves well if they can understand, at least at a high level, the type and nature of data sources being leveraged to render an AI decision. Companies should work with their technology providers to receive a manifest of data sources being leveraged by the provider in their AI and ML models and maintain this dictionary and taxonomy within their organization.

Technology Best Practice 3 - Automated Testing for Inadvertent Bias and Proxy Discrimination

One of the challenges that companies struggle with is the ability to leverage technology in order to prove absence of inadvertent bias or proxy discrimination in AI models in an automated manner.

Some "early adopters" have established and robust auditing procedures to be able to do so, but lack automated technological capabilities to be able to clearly prove that AI does not favor one decision over the other. In addition to existing policies and processes that are currently employed by companies, firms might consider exploration of emerging practices such as the Bayesian Improved Surname and Geocoding (BISG) methodology. Developed by the

RAND Corporation, BISG can help organizations produce accurate and cost-effective estimates of racial and ethnic disparities within their data sets, and in doing so highlight areas where improvement might be needed. Most companies intentionally never collect race and ethnicity information on applicants expressly for the purpose of feeding an AI model to generate an output. The BISG method uses indirect estimation in order to generate an estimate of race and ethnicity when these data elements are unavailable. By combining geocoded address and last name, and leveraging Census Data, the BISG algorithm can predict race and ethnic probability of an individual. Companies can potentially use this method as part of their auditing process to estimate the racial and ethnic composition of a customer base (post decision), and compare their AI-based adjudications for these groups. It should be noted that this practice is still quite new, and as such companies should proceed with patience as regulatory bodies vet and issues guidelines around how the BISG could be used within an organization.

Technology Best Practice 4 - Automated QC Processes

Companies undertake a trusted partner approach to their technology providers, wherein little to no quality assurance and quality control is conducted at the destination (when the data is received).

Since these technology providers are mature in their practices, companies generally take it for granted that the data being sent to them is thoroughly vetted, tested, is of high quality, and is robust for consumption in AI and ML models. While this practice has worked for the past several years, as firms have been ingesting structured data, with the potential use of unstructured and semi-structured data in the future, it will be important for Chief Data Officers and Chief Information Officers to establish robust automated quality control processes to benchmark, measure, and highlight data quality across these inbound data sets.

Technology Best Practice 5 - APIs and Cloud as Core Requirements

The growth of AI has been directly proportional to the increasing popularity of cloud-based systems and the availability of APIs to facilitate integrations between companies and systems provided by technology providers.

Although there is no implied causality between the two, it should be noted that many industries are rapidly embracing an API-driven ecosystem, allowing companies to connect with data on customers as well as third-party providers. It was evident through the COVID-19 pandemic that companies who had embraced a cloud-first ideology fared better through the pandemic than those who did not.

CIOs should continue to consider a cloud-first mindset, and migrate to the cloud where it makes sense. The cloud can be cost prohibitive to smaller firms, where the volume and scale would not make sense to move everything that a firm has into the cloud. The API ecosystem allows companies and providers to exchange information in a secure manner, and companies that are devoid of a cloud strategy or an ability to consume and publish APIs could find themselves at a significant disadvantage in the future. These firms will find themselves investing a lot more in doing remedial integration activities to be able to exchange information. The time-to-market that AI-based decisions inspire might be tempered if organizations do not have the technical capabilities with APIs and cloud to enable desired business speeds.

SECTION THREE

AI BEST PRACTICES

PART C – EXPLAINABLE AI

Chapter Twenty-Six: Fairness and Transparency in AI

The potential for AI's transformative abilities is limitless, and while AI is likely to fundamentally transform entire industries, absent appropriate guardrails, it also carries the threat of enabling, reinforcing, and exacerbating societal biases. Bias in AI refers to the systematic and unfair discrimination against certain individuals, or groups of individuals, by an AI algorithm, based on attributes such as race, gender, age, sexual identity, sexual orientation, disabilities, socioeconomic status, etc. One of the primary concerns across the sprawling field of AI, regardless of type of implementation – Computer Vision, Speech to Text, Machine Learning, etc. - is always appropriately focused on the topic of avoiding unintentional bias and proxy discrimination in data and AI. This is reflective of the fact that industries, despite the absence of regulation, take ethical and fairness very seriously, and prioritize these concerns over methodological ones. Bias in AI can be exhibited in one of three ways as illustrated in Figure 52:

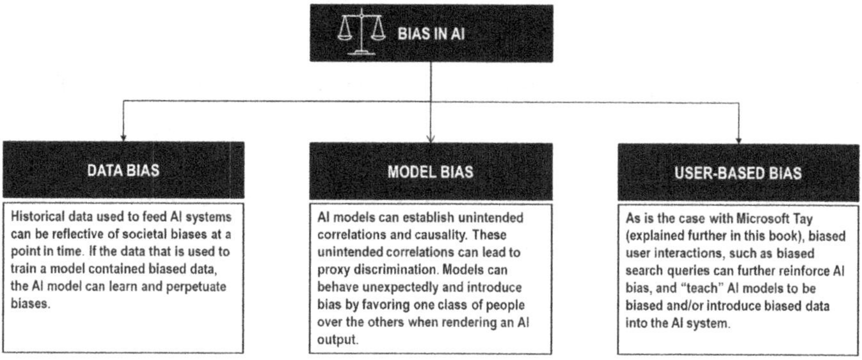

Figure 52: Causes of Bias in AI systems

The greater the volume of data and the more complex a model, the greater is the opportunity for correlation and pattern inference. It is impossible for humans to be able to analyze and discern correlations between traits that an AI model learns in contrast to what it was taught to learn. AI models "self-learn" by crunching through vast amounts of data and by looking for correlations within this data. Sometimes, AI systems can inadvertently correlate disparate elements of data to arrive at a prediction or recommendation. Although this correlation is unintended, it can have far reaching impacts in terms of its ultimate output. Consider these two concrete examples:

i. Amazon.com and Recruiting:

Starting in 2014, one of Amazon's machine learning teams had been developing a program to vet the resumes of job applicants. The goal was to vet the resumes in an automated manner, reduce reliance on human recruiters, and provide top talent recommendations to these human recruiters, allowing them to recover some of their capacity – while delegating some of the rote, manual work to the algorithms. Much like shoppers on the e-commerce online retailer's platform are allowed to rate products on a one through five scale, this recruitment screening AI tool scored job candidates on a five-scale.

According to a Reuter's article describing the issue, "Automation has been key to Amazon's e-commerce dominance, be it inside warehouses or driving pricing decisions" (Dastin, 2018). The article goes on to quote an Amazon employee who stated that the company "wanted it to be an engine where I'm going to give you 100 resumes, it will spit out the top five, and we'll hire those." After about a year of the AI being operational, Amazon realized that the system had not been rating candidates for software developer and IT jobs in a gender-neutral manner.

The AI/ML models that had been trained to evaluate candidates by pattern recognition and observations in resumes that were submitted to Amazon over a decade. Reflecting that these professions have historically been male dominated, the model self-learned that male candidates were "preferable." The model penalized applicant resumes that included words such as "women's" as the word appeared in terms such as "women's chess club champion," downgraded graduates of two all-women colleges, etc. While Amazon "did not dispute that recruiters looked at the recommendations generated by the recruiting engine" (Dastin, 2018), the company does claim that their recruiters did not leverage the algorithm to evaluate candidates or make hiring decisions. The Amazon story has been a lesson to other organizations such as Goldman Sachs, and Hilton Worldwide Holdings, which had been planning to leverage similar utilities for their own recruitment practices.

ii. Microsoft's Chatbot:

The second example is technology titan, Microsoft, and their experiment with Microsoft Tay. Microsoft released Tay (an acronym for "Thinking About You") on March 23rd, 2016. Tay was an AI chatbot that was designed to self-learn from the internet, emulate an American teenage girl, and automatically post messages to social media site Twitter in the form of "tweets" under the Twitter handle "TayTweets." Tay was devoid of human oversight given the "self-learning and tweet post" nature of its purpose. In less than 16 hours, Tay self-learned from the internet to become extremely racist, xenophobic, and sexist – these learnings manifesting themselves in the AI's tweets. Microsoft had to take Tay offline and publicly apologize.

Amazon and Microsoft are both highly sophisticated technology titans. If these missteps could befall these digitally native organizations, consider the challenge posed by "guardrails-free" AI use across sectors with anachronistic technologies.

Understanding the Need for AI Transparency and Fairness

If humans are unable to understand how an AI arrived at a conclusion or recommendation that it did, it becomes very difficult to trust in the decision. Regulated industries such as healthcare and insurance must be particularly watchful to ensure they are complying with emerging regulations intended to protect the consumer. The use of Generative AI and the implications of cheating in academia is one such concern with ethics and AI, but there are very real concerns that AI might make decisions that are inadvertently biased or accidentally discriminatory in nature. These kinds of challenges can occur due to the inherent quality of data being used, or by the underlying AI and machine learning (ML) models themselves.

Federal law explicitly prohibits discrimination against groups of people who share certain protected characteristics including age, gender, gender identity, race, color, religion, national origin, sexual orientation, disabilities, veteran status, etc. Companies scrutinize their AI models closely for concerns that these models could introduce inadvertent and unintentional biases that result in adverse decisions for customers. While organizations can have time-tested and robust audit procedures established for their traditional, human processes along their value chain, AI-based decisions can seemingly be a black box to those who are not directly involved with them. In some respects, there is a healthy fear of the unknown at play with decision-making shifting from the judgement, skill, experience, and ethical behaviors of humans to AI. Mitigating bias and proxy discrimination in AI is not a singular task that can happen in isolation. It is an ongoing, multi-faceted effort that requires focus, prioritization, and commitment.

Additionally, if an AI model has been provided instructions to disregard specific data elements or attributes, the model will always disregard these traits. However, if the model does not have just as specific instructions to ignore certain other related traits, the model might interpret these other traits that result in unintentional bias. AI models can develop unintentional bias either by the accidental reinforcement of historical discriminatory patterns, or via discovering new discriminatory patterns via unexpected correlations.

Of these two, it is easier to identify and rectify historical discriminatory patterns, whereas it is more challenging to identify unexpected correlation since AI models consume vast amounts of data to establish connections, and could and accidentally correlate unexpected data attributes. Consider a hypothetical scenario where students across the United States are applying for a prestigious Ivy League university. Imagine that this determination was being driven by an AI model, and that this AI model leveraged geographical location as a key input. In such a circumstance, it is entirely foreseeable that the zip code/geographic location of the student applicant is highly correlated with ethnicity. In other words, zip codes can reflect majorities of races that reside within that zip code. Therefore, if the student's location is highly correlated with ethnicity, it is likely that the AI model would inadvertently favor certain ethnicities over others.

Mitigating Bias

The AIM Framework© strongly recommends that organizations not wait until there are regulatory frameworks and guidelines governing mitigating bias in AI to institute mechanisms to identify and remediate bias in AI. Mitigating bias needs to be second nature, and central to your enterprise AI strategy. Companies have to be proactive to ensure that their AI systems are transparent and free of bias. My guidance to companies is that your company should make news, not be the news, and unless proactive measures are enacted to intercept and remediate potential bias in AI, you are more than likely to be the news. These types of issues can be fatal to your enterprise AI program, shatter stakeholder trust, and invite ongoing scrutiny across all your technology, not just your AI program. The good news is that there are several ways that organizations can institutionalize mitigating bias in AI within their enterprise AI programs. Figure 53 outlines five of the aspects – at a very high level – that your firm might consider focusing on immediately.

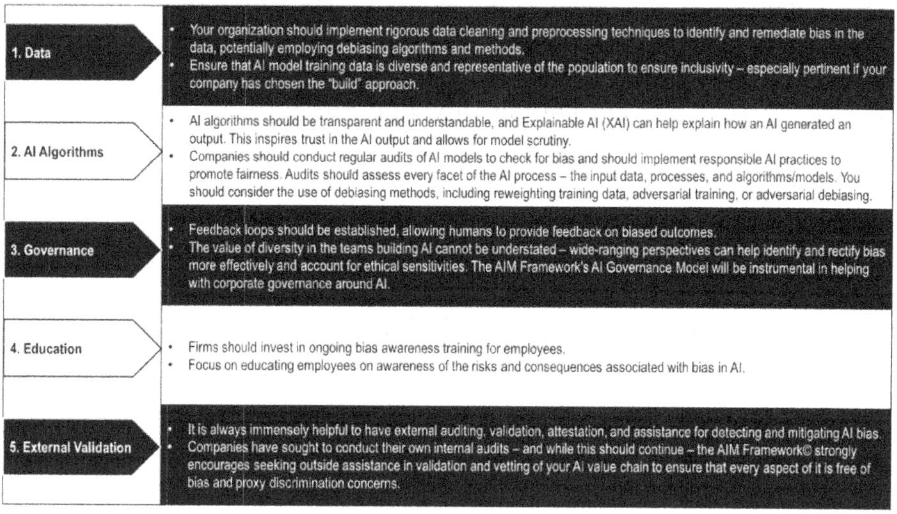

Figure 53: Methodologies to detect, mitigate, and remediate bias in AI

One of the prominent concerns regarding AI as pertains to proxy discrimination is that an AI algorithm will execute on instructions that it has been supplied. If an AI model has been provided instructions to disregard specific attributes, the model will always disregard these attributes, but if the model does not have just as specific instructions to ignore certain other related attributes, the model might interpret these other attributes that result in unintentional bias. The greater the volume of data and the more complex a model, the greater is the opportunity for correlation and pattern inference. It is impossible for humans to be able to analyze and discern correlations between attributes that an *AI model learns* in contrast to *what it was taught to learn*. The practice of micro-segmentation, and datasets that are outdated, too simple, or contain unrelated attributes, are two areas that their firms scrutinize closely to ensure they are mitigating any opportunities for bias or proxy discrimination to infiltrate their AI practices. In addition to the challenge of attribute correlation, micro-segmentation, a practice used in AI model development, can cause challenges. Micro-segmentation categorizes and subdivides data into smaller groups based on common attributes. Micro-segmentation, when dealing with large data sets, and low explainability from a multitude of variables can make it challenging to understand an AI model's decision-making.

An illustrative example, to continue the examination of a use case within the life insurance industry (as cited in Enterprise Best Practice 5 – Regulatory Landscape), comes from the National Association of Insurance Commissioners (NAIC). The NAIC's Automated and Accelerated Underwriting Working Group, which drafts recommendations/regulatory frameworks for AI in life insurance underwriting, summarizes what life insurance companies should be expected to do to ensure their models are free of inadvertent bias and proxy discrimination. These draft guidelines are good examples for your organization, regardless of industry and sector to evaluate and extrapolate for your needs. The guidelines state that: "Insurers and other parties involved in accelerated underwriting in life insurance should:

i. Take steps to ensure data inputs are transparent, accurate, reliable, and the data itself does not have any unfair bias.
ii. Ensure that the use of external data sources, algorithms or predictive models are based on sound actuarial principles with a valid explanation or rationale for any claimed correlation or causal connection.
iii. Ensure that the predictive models or machine learning algorithm within accelerated underwriting has an intended outcome and that outcome is being achieved.
iv. Ensure that the predictive models or machine learning algorithm achieve an outcome that is not unfairly discriminatory.
v. Be able to provide the reason(s) for an adverse underwriting decision, whether the decision is based on data subject to FCRA or not, to the consumer and all information upon which the insurer based its adverse underwriting decision.
vi. Take steps to protect consumer privacy and ensure consumer data is secure.
vii. Have a mechanism in place to correct mistakes if found.
viii. Produce information upon request as part of regular filing submissions reviews or market conduct examinations" (NAIC, 2022).

The adoption of AI will continue to expand and mature across every industry. As these implementations continue to mature, companies should continue to prioritize the sensitive topics of inadvertent bias and proxy discrimination. While guidelines - both from consultants, and regulators, continue to develop - with regulatory bodies recognizing the need for urgency in developing and issuing these guidelines, companies have done well within their own companies to manage their own AI implementations, seeking the highest standards of fairness, ethics, and transparency.

Although the best practice recommendations outlined in this AIM Framework© – from the ten enterprise best practices to the ones specific to the People, Process, and Technology domains – will not definitively mitigate and eradicate inadvertent bias and proxy discrimination, following industry best practices will certainly equip and enable companies to identify and eliminate it. The AIM Framework© strongly recommends that companies continue to be proactive in their approach.

Chapter Twenty-Seven: Explainable AI (XAI)

Ensuring that AI models allow for transparency such that humans can clearly understand how the AI arrived at the decision or the recommendations that it did is the central tenet of a concept known as Explainable AI (XAI). The tenet of explainability – being able to explain how and why an AI and ML model made the decision that it did – is vital. This chapter brings together a plethora of published works on XAI, to provide an overview of XAI from experts in the field.

The Concept of XAI

The term Explainable AI (XAI) was first coined by the Defense Advanced Research Projects Agency (DARPA) as a part of several research initiatives designed to address a critical shortcoming of AI at scale at the time – the ability for people to clearly understand why an AI system arrived at the decision it did, with the data it learned from, and to be able to methodically track this decision-making. DARPA's investigation into XAI was initiated due to the basic premise that the greater complexity within an AI and ML model, the more challenging it becomes to interpret the result and understand how the result was achieved. According to an article that appeared in the MIT Technology Review, "The Defense Advanced Research Projects Agency (DARPA), a division of the Defense Department that explores new technologies, is funding several projects that aim to make artificial intelligence explain itself. The approaches range from adding further machine-learning systems geared toward providing an explanation, to the development of new machine-learning approaches that incorporate an elucidation by design."

A MIT Technology Review article, quoting the DARPA Program Manager, states: """We now have this real explosion of AI," says David Gunning, the DARPA program manager who is funding an effort to develop AI techniques that include some explanation of their reasoning. "The reason for that is mainly machine learning, and deep learning in particular."" (Knight, 2017). In a study covering DARPA's approach to XAI, Dr. Matt Turek states "Dramatic success in machine learning has led to a torrent of Artificial Intelligence (AI) applications. Continued advances promise to produce autonomous systems that will perceive, learn, decide, and act on their own. However, the effectiveness of these systems is limited by the machine's current inability to explain their decisions and actions to human users. The Department of Defense (DoD) is facing challenges that demand more intelligent, autonomous, and symbiotic systems. Explainable AI - especially explainable machine learning - will be essential if future warfighters are to understand, appropriately trust, and effectively manage an emerging generation of artificially intelligent machine partners.

The Explainable AI (XAI) program aims to create a suite of machine learning techniques that: Produce more explainable models, while maintaining a high level of learning performance (prediction accuracy); and enable human users to understand, appropriately trust, and effectively manage the emerging generation of artificially intelligent partners. New machine-learning systems will have the ability to explain their rationale, characterize their strengths and weaknesses, and convey an understanding of how they will behave in the future. The strategy for achieving that goal is to develop new or modified machine-learning techniques that will produce more explainable models. These models will be combined with state-of-the-art human-computer interface techniques capable of translating models into understandable and useful explanation dialogues for the end user. Our strategy is to pursue a variety of techniques in order to generate a portfolio of methods that will provide future developers with a range of design options covering the performance-versus-explainability trade space" (Turek, 2017).

In his article, Dr. Turek presents a simple concept for XAI that is based on finding answers to questions that AI devoid of explainability cannot address. These questions that XAI solves for, as described by Dr. Turek, are as follow:

From: "Why did you do that?" – To: "I understand why."

From: "Why not something else?" – To: "I understand why not."

From: "When do you succeed?" – To: "I know when you succeed."

From: “Why do you fail?” – To: “I know when you fail.”

From: “When can I trust you?” – To: “I know when to trust you.”

From: “How do I correct an error?” – To: “I know why you erred.”

IBM explains XAI as “Explainable artificial intelligence (XAI) is a set of processes and methods that allows human users to comprehend and trust the results and output created by machine learning algorithms. Explainable AI is used to describe an AI model, its expected impact, and potential biases. It helps characterize model accuracy, fairness, transparency, and outcomes in AI-powered decision making. Explainable AI is crucial for an organization in building trust and confidence when putting AI models into production. AI explainability also helps an organization adopt a responsible approach to AI development. As AI becomes more advanced, humans are challenged to comprehend and retrace how the algorithm came to a result. The whole calculation process is turned into what is commonly referred to as a “black box" that is impossible to interpret. These black box models are created directly from the data. And, not even the engineers or data scientists who create the algorithm can understand or explain what exactly is happening inside them or how the AI algorithm arrived at a specific result.” (IBM, n.d.).

Fundamentals and Value of XAI across Industries

A comprehensive paper by McKinsey & Company published in 2022 provides a fairly detailed examination of the need for XAI across industries. The same article, defining XAI as "Explainability is the capacity to express why an AI system reached a particular decision, recommendation, or prediction" (Grennan, Kremer, Singla, & Zipparo, 2022), also provides guidance on how businesses can meet this need.

Espousing the value of keeping XAI as one of the central tenets to AI and ML development, the article states that "As artificial intelligence informs more decisions, companies' AI systems must be understood by users and those affected by AI use." The study elucidates that the concept of XAI is becoming increasingly business critical as AI continues to become ubiquitous across an enterprise, and it is vital to understand how any particular model arrived at a particular conclusion. It is just as important for an organization to clearly understand the underlying data that was used to arrive at that particular conclusion. Reliance on AI to drive business is one thing, but it is more critical for a business to understand that this reliance is not unfounded and that AI-rendered recommendations can be trusted. This trust is at the core of XAI and the study states that while getting XAI right is vital, XAI also translates into direct positive business outcomes.

Research conducted by McKinsey that is cited in this study states, "companies seeing the biggest bottom-line returns from AI - those that attribute at least 20 percent of EBIT to their use of AI - are more likely than others to follow best practices that enable explainability. Further, organizations that establish digital trust among consumers through practices such as making AI explainable are more likely to see their annual revenue and EBIT grow at rates of 10 percent or more." The McKinsey paper provides five ways that XAI can benefit an organization, citing increased productivity, building trust in AI and adoption of AI, providing new and previously unknown business insights, ensuring AI provides business value, and mitigation of regulatory and other risks.

Dr. Hugh Watson, professor at the University of Georgia, in his paper entitled "The Need for Explainable Processes and Algorithms" (Watson, 2020), which espouses the need for XAI, emphasizes the changing regulatory and compliance landscape, specifically calling attention to the myriad of data privacy regulations, such as GDPR in Europe, CCPA/CPRA in California, etc. Dr. Watson highlights the vital nature for XAI not only from a regulatory and compliance perspective, but from a consumer point of view, stating "People have relatively little interest or concern about non-threatening applications such as online product recommendations or AI-powered chatbots, but they do care about: i. A prediction that a person is likely to commit another crime and is given a prison sentence rather than probation, ii. The decision whether a person is given insurance and at what cost is based on an algorithm, iii. A prediction that a person would be a good employee and is granted an interview based on an algorithm's analysis of the person's resume, iv. A prediction that an individual isn't a good credit risk and is denied a loan, v. A treatment plan created based on an algorithm's analysis of a patient's test results and symptoms. People want to know that models are fair, trustworthy, secure, and explainable and that remediation processes are in place when errors are made."

Human-Centered and Trustworthy AI

Trust, specifically trust in the decision that an AI engine has rendered, is a recurring theme across several studies that have been conducted on XAI. It is apparent that it is impossible to detach XAI from being inherently human-centric. This is simply because the target beneficiaries of explainability and the building of trust in AI, are humans. For humans to embrace broad adoption of AI, to have confidence in an algorithm's recommendations, and to be able to communicate the decision-making process to each other, it is imperative that AI inspires trust across the stakeholder spectrum.

A study by the National Institute of Standards and Technology (NIST) describes the concept of the need for AI to build trust and confidence by stating that "Explainable AI is one of several properties that characterize trust in AI systems. Other properties include resiliency, reliability, bias, and accountability. Usually, these terms are not defined in isolation, but as a part or set of principles or pillars. The definitions vary by author, and they focus on the norms that society expects AI systems to follow. Based on the calls for explainable systems, it can be assumed that the failure to articulate the rationale for an answer can affect the level of trust users will grant that system. Suspicions that the system is biased or unfair can raise concerns about harm to individuals and to society. This may slow societal acceptance and adoption of the technology" (Phillips, Hahn, Fontana, Yates, & Greene, 2021).

In their research entitled "A historical perspective of explainable Artificial Intelligence" (Confalonieri, Coba, Wagner, & Besold, 2020), the authors provided a historical context for the genesis of XAI, how it is understood today, and what it might look like in the future. This paper, which explores the concept of explainability from a technical perspective in depth (outside the scope of this particular research), presents XAI explainability criteria that the authors posit would serve as core facets in the growth and development of human centered XAI. These recommended criteria include:

i. Causal (understanding relationships between inputs to a model and commensurate outputs)
ii. Counterfactual (not just understanding why an event, event x, happened, but moreover understanding why event x happened versus event y)

iii. Social (being able to explain AI/ML decisions geared towards a specific human – how that particular person prefers to learn and understand)

iv. Selective (being able to explain AI/ML decisions geared towards what a specific human in a specific role as a stakeholder needs to know)

v. Transparent (being able to help humans understand how a particular decision was reached and why that decision was reached, while being able to protect exposing the model or training data and being able to balance transparency and privacy)

vi. Semantic (being able to support common sense reasoning and convey the same information personalized to disparate constituencies of stakeholders)

vii. Interactive (being able to allow feedback to be incorporated into further model and AI refinement).

Belle and Papantonis, in their paper entitled "Principles and Practice of Explainable Machine Learning" (Belle & Papantonis, 2021), describe what various stakeholders might be concerned about with XAI in the AI value chain, along with how they would benefit from XAI. The authors state that a data scientist in the value chain might be concerned with understanding the model, debugging the model, and improving model performance; a business owner would be concerned with the need to understand the model, evaluate fit for purpose, and agree for use; a model risk stakeholder would be concerned with the need to challenge the model, ensure model robustness, and approve of it; a regulator would be concerned with checking the model's impact on consumers, and verification of reliability; and the end consumer would be concerned with answering what impact the model has on them, and what actions can the consumer take in response to the model's decisions.

Thematically resonant with the NIST standards outlined above, Belle and Papantonis out line six key areas that XAI, specifically XAI with human stakeholders along the value chain, should keep in mind: "a. Correctness: Are we confident all and only the variables of interest contributed to our decision? Are we confident spurious patterns and correlations were eliminated in our outcome? b. Robustness: Are we confident that the model is not susceptible to minor perturbations, but if it is, is that justified for the outcome? In the presence of missing or noisy data, are we confident the model does not misbehave? c. Bias: Are we aware of any data-specific biases that unfairly penalize groups of individuals, and if yes, can we detect and correct them? d. Improvement: In what concrete way can the prediction model be improved? What effect would additional training data or an enhanced feature space have? f. Transferability: In what concrete way can the prediction model for one application domain be applied to another application domain? What properties of the data and model would have to be adapted for this transferability? g. Human comprehensibility: Are we able to explain the model's algorithmic machinery to an expert? Perhaps even a lay person? Is that a factor for deploying the model more widely?"

This study outlines various approaches on XAI, criteria for how one can evaluate explainability as a concept, and the type of explanations that one might expect. The paper takes an in-depth look at the notion of transparency in AI and ML, which seems to echo the central predicate of explainable AI. The authors describe various explanation types inherent to XAI, and state that while there is overlap across the various explanation types that have been outlined in the study, explanation types are segmented such that each of them address a different question along the XAI value chain. The authors posit that these resulting approaches of segmenting explanation types for responding to specific questions in the value chain allow us to leverage a model's unique features to produce explanations and improve the fidelity of the model, as well as allow exploration of a model's internal working. They recommend that XAI would benefit by focusing on explaining how a model works (causal analysis), rather than only explaining the outcome and the decision resulting from a model. The authors state that as XAI continues to mature, the concepts of causal analysis should be featured prominently into XAI, since causal analysis is already a major driver in intrinsic problems in other facets of AI, such as addressing the concepts fairness and bias in ML. The authors expect causal analysis to play an integral part in the future of XAI literature.

Challenges in XAI Implementation

The authors of the McKinsey study are transparent about the quest for XAI being a challenging one for organizations to tackle for several reasons. A prominent challenge facing XAI is the sheer complexity of today's AI and ML models. The inherent sophistication and complexity of these models, such as deep learning and neural networks, pose a challenge for humans to understand, let alone to be able to explain the decision-making of these models. XAI, at its very basic, requires an understanding of the AI model and the underlying data that was used to train the AI model. As the complexity of the model increases, it becomes increasingly difficult to identify how a specific decision was reached, since these models learn over time. Models can deliver outcomes and decisions within fractions of a second, ingesting a massive amount of data, determining the predictive power of multiple algorithmic permutations and combinations, and updating the model itself at these speeds. It can be relatively straight-forward to discern causality and explain a point-to-point (A to B) decision, but this becomes an order of magnitude more difficult as models repeatedly interpolate massive amounts of data.

A variety of advanced ML engines along the IT Software Supply Chain can exponentially complicate this challenge. AI and ML engines can often be treated as “black boxes” and the way to solve for XAI is not simply for humans to be able to explain how an AI/ML system operates, but to develop automated processes and technologies that can help experts understand the models, such that they can then explain it to others. The authors of the McKinsey study recommend establishment of governance frameworks, having the appropriate processes, and leveraging the right tools and technologies. XAI also has diverse interests across the stakeholder spectrum as distinct AI consumers across an enterprise value chain might have differing vested interests in what they expect from explainability.

As the McKinsey article notes, “A bank that uses an AI engine to support credit decisions will need to provide consumers who are denied a loan with a reason for that outcome. Loan officers and AI practitioners might need even more granular information to help them understand the risk factors and weightings used in rendering the decision to ensure the model is tuned optimally. And the risk function or diversity office may need to confirm that the data used in the AI engine are not biased against certain applicants. Regulators and other stakeholders also will have specific needs and interests.” (Grennan, Kremer, Singla, & Zipparo, 2022).

Andrea Brennen explores this multifaceted need for XAI to be relevant and pertinent to a broad spectrum of stakeholders in a qualitative research paper entitled “What Do People Really Want When They Say They Want Explainable AI?” (Brennen, 2020). This qualitative research, which interviewed 40 stakeholders and conducted two focus groups for a total of 60 stakeholders across a span of nine months, sought to understand how disparate stakeholders understand XAI. Intended to support a broad overview of the problem of XAI, this research has provided two notable findings that state: “(1) current discourse on Explainable AI is hindered by a lack of consistent terminology; and (2) there are multiple distinct use cases for Explainable AI, including: debugging models, understanding bias, and building trust. These uses cases assume different user personas, will likely require different explanation strategies, and are not evenly addressed by current XAI tools.” Another challenge for XAI adoption might be the lack of a shared understanding. Underscoring the point of the lack of consistent language and taxonomy of what XAI means, Andrea Brennan enumerates synonyms that stakeholders in her study group used for “Explainable” in context of AI. These terms include “Accountable, Auditable, Certifiable, Fair, Inspectable, Interpretable, Justifiable, Operational, Ready-to-Use, Reliable, Repeatable, Reproducible, Responsible, Self-service, Tested, Transparent, Trusted, Unbiased, Understandable, and Verifiable.” Across the spectrum of stakeholders studied for this research,

Andrea Brennen discovered that various stakeholders often have multiple and overlapping reasons for having a vested interest in XAI. These interests include the need to build trust, ensure models are not treated as a black box, for debugging models, and identifying bias.

In a comprehensive paper on XAI, "Explainable Artificial Intelligence (XAI): Concepts, taxonomies, opportunities and challenges toward responsible AI" (Arrietaa, et al., 2020), the authors summarize the tenets of XAI as: Understandability, Comprehensibility, Interpretability, Explainability, and Transparency. This paper, which introduces the concept of "Responsible AI," defines XAI as "Given an audience, an explainable Artificial Intelligence is one that produces details or reasons to make its functioning clear or easy to understand." The authors encapsulate the end goal of XAI with their definition of Responsible AI, stating that it is "a methodology for the large-scale implementation of AI methods in real organizations with fairness, model explainability and accountability at its core. Our ultimate goal is to provide newcomers to the field of XAI with a thorough taxonomy that can serve as reference material in order to stimulate future research advances, but also to encourage experts and professionals from other disciplines to embrace the benefits of AI in their activity sectors, without any prior bias for its lack of interpretability."

Effectuating XAI

In a paper published in 2021, the National Institutes of Standards and Technology (NIST), which outlined Four Basic Principles of XAI, echoed elements of the need to ensure that any XAI implementations should seek to ensure that the entire spectrum of stakeholders and their vested interests are considered. The authors provide an example in orienting their XAI basic principles around the different needs of different stakeholders in the value chain by stating: "AI developers and designers may have very different explanation needs than policy makers and end users. Therefore, why an explanation is requested and how it is delivered may differ depending on the AI users.

These four principles are heavily influenced by considering the AI system's interaction with the human recipient of the information. The requirements of the given situation, the task at hand, and the consumer will all influence the type of explanation deemed appropriate for the situation. These situations can include, but are not limited to, regulatory and legal requirements, quality control of an AI system, and customer relations. Our four principles of explainable AI systems are intended to capture a broad set of motivations, reasons, and perspectives. The principles allow for defining the contextual factors to consider for an explanation, and pave the way forward to measuring explanation quality." (Phillips, Hahn, Fontana, Yates, & Greene, 2021).

Two of the recommendations in the McKinsey paper for businesses to make XAI possible directly overlap some of the best practice recommendations outlined in Chapter Five. The McKinsey paper recommends that companies "Establish an AI governance committee to guide AI development teams." This directly correlates to the People Dimension Best Practice of "AI and ML Governance Model" that the AIM Framework© recommends. The second recommendation of this paper is for firms to "Invest in the right talent, explainability technology, research, and training," which matches up to "Enterprise Best Practice 7: Focus on Data Quality and Data Literacy," "People Dimension Best Practice 5: Knowledge Sharing across Enterprise," and "Process Dimension Best Practice 1: Processes that Promote Transparency and Explainability."

Similarly, Dr. Watson offers his own four steps that companies should take in the pursuit of XAI, stating that "In addition to the use of technology, organizational actions in governance and model development processes can help ensure that analytical applications meet the demands of management, laws and regulators, and calls by the public for greater explainability" (Watson, 2020). Dr. Watson's four recommended steps are: "1. Understand how important explainability is in your industry and specific applications, 2. Build explainability into the entire model development process, 3. Satisfy the legal requirements for explainability, and 4. Expand governance to include explainability." Dr. Watson in his paper for the Business Intelligence Journal outlines three modeling and design approaches to enable XAI – Deep Explanation, New Algorithms, and Model Induction. Dr. Watson also states that in addition to these three design approaches, firms are free to create their own algorithms to ensure they are solving for XAI bespoke to their specific needs. "Companies that need explainability may create their own algorithms to satisfy specific needs or requirements. For example, the lead-scoring function in Salesforce's Sales Cloud Einstein provides insights into how particular leads are scored, which is important to sales teams. To meet the requirements of the Fair Credit Reporting Act, Equifax must be able to tell consumers the top four reasons why they did not get a perfect credit score and give reasonable recommendations (i.e., remediation) for how to improve their score. In response, Equifax created and patented NeuroDecision to meet these regulatory requirements for explainable credit scoring." This is valuable for all companies seeking their own implementations of XAI to keep in mind.

In addition to the aforementioned papers on how XAI can be effectuated across enterprises, the National Institutes of Standards and Technology (NIST), in September 2021, outlined their Four Basic Principles of XAI. The paper states that NIST proposes that "explainable AI systems deliver accompanying evidence or reasons for outcomes and processes; provide explanations that are understandable to individual users; provide explanations that correctly reflect the system's process for generating the output; and that a system only operates under conditions for which it was designed and when it reaches sufficient confidence in its output" (Phillips, Hahn, Fontana, Yates, & Greene, 2021). This study defines its four basic principles as: i. Explanation, which states that an AI system deliver or contain accompanying evidence to its commensurate results and/or processes, ii. Meaningful, which states that an AI system needs to provide explanations that are meaningful and comprehensible to the target stakeholder/s, iii. Explanation Accuracy, which means that the explanation needs to accurately reflect that systems processes and/or be an accurate reflection of the rationale behind why a particular decision was reached, and iv. Knowledge Limits, which means that an AI system should only operate under the conditions it was designed for and should only execute when an adequate/appropriate level of confidence in its decision-making ability has been attained.

There are several established papers on the concept of Explainable AI (XAI). It is recommended that practitioners augment the Industry Best Practices recommended in this research with further reading and an understanding of the concept of XAI and how they can align their AI and ML practices to XAI recommended best practices.

SECTION THREE

AI BEST PRACTICES

PART D – DATA, DATA, DATA

Chapter Twenty-Eight: Data, Data, Data – Part 1

The story of AI can only be written through the words of data.

How many of us can define data? According to the New Oxford American Dictionary, data is defined as "Facts and statistics collected together for reference or analysis" (Oxford Dictionaries, 2011), while the American Society for Quality (ASQ) defines data as "A set of collected facts. There are two basic kinds of numerical data: measured or variable data, such as '16 ounces', '4 miles' and '0.75 inches'; and counted or attribute data, such as '162 defects'" (ASQ, n.d.). The criticality of data for the success of your enterprise AI program has continually been reiterated throughout this book. Your organization can have sophisticated AI models – whether via a "build" or a "buy" approach. These AI models might be created by sophisticated AI algorithms that are programmed using the best, most performant, secure code. However, having the best AI models without appropriate data is amenable to having a terrific car without the fuel to power it. You won't be able to get anywhere in either case.

An interesting aspect about the "Age of AI" is that organizations that had finally come around to taking their data assets seriously -

recognizing the untapped potential of their "data gold mines" - were starting to invest in their data strategy and management programs in earnest. As they were finally starting to firm up the execution of their data and analytics strategies, the exponentially growing "Age of AI" burst onto the scene, and has exerted unique pressures. These firms have suddenly found themselves underprepared for managing AI programs atop their fledgling data programs. It bears repeating: "Great AI + Bad Data = Terrible AI." Attempting to establish a top-notch AI program without a strong data foundation is akin to establishing a skyscraper on a leach field. You will be unable to erect a magnificent AI edifice absent a strong data foundation.

Data is the lifeblood of your organization, regardless of if this data is being leveraged for AI, analytics, operational reporting, or to develop management dashboards. Data, when used correctly, can directly impact your profitability, help to streamline your operations, and enhance customer experience. In addition to AI, data plays a key role in digitization efforts, in the so-called "data-driven digital transformations." As companies continue to understand and leverage the power of data, it is critical that they also implement strategies to run these data-related programs effectively. This chapter will refrain from delving into data strategy and governance in depth. What this chapter will seek to do is to inspire the importance – at an adequately high-level – of data within the context of an AI strategy.

Data - So. Much. Data

It is said that we are drowning in data, but thirsty for knowledge. As astonishing as it is to witness the rapid evolution and advancement of AI, the sheer order of magnitude growth of data on a daily basis is not something we think of, but is just as – if not more – awe inspiring.

The total amount of data consumed globally at the start of the 2020s, was 79 zettabytes. This number will grow to over 180 zettabytes by 2025. One zettabyte is equal to a thousand exabytes, a billion terabytes, or a trillion gigabytes. One zettabyte is roughly equal to 1125899906842600 megabytes (MB). 79 zettabytes equate to 88946092640567000 megabytes (MB) - or 8.8946092640567 × 1016 megabyte in scientific notation. Consider how unimaginably large a zettabyte is. Megabytes, gigabytes, and terabytes seem quite arbitrary to most people, so let's use time as an analogy.

A single day on Earth, 24 hours, converts to 86,400 seconds.

A million seconds is 12 days.

A billion seconds is approximately 32 years.

A trillion seconds is approximately 32,000 years.

Imagine that 1 byte is 1 second, and you can begin to fathom the incredibly large amounts of data we are producing.

Digital photos are a good proxy to try and comprehend this growth. The first smartphone with a digital camera, the Sharp Electronics J-SH04 J-Phone, was introduced in November of 2000 (or, according to some, the Samsung SCH-V200 that was introduced 5 months prior). This was the first phone that allowed users to share photos with each other without first needing to connect this phone to a computer and transfer the photos. There had been nearly 7.9 trillion photos stored on computer drives, PCs, mobile devices, in the cloud, printed out, etc. in the start of the 2020s. Assuming you stored one photo continually per second, it would take you approximately 250,000 years to reach that number. This number grew to approximately 10 trillion photos in 2023. As a frame of reference, 250,000 years ago, the greatest technological advancement on this planet was when Neanderthals mastered fire.

Explaining Data Governance in the Age of AI

Data and information are invaluable enterprise assets in the 21st century and the so-called "Fourth Industrial Revolution" that we are currently in. For a majority of established and mature industries, data has never historically been treated as a "first class citizen," wherein there has been a lack of understanding that data is and can serve as an enterprise asset. Several mature organizations are still attempting to reconcile the vast treasure trove of data that they are in custody of with the need of the organization to collectively rally a cultural shift as to how data is treated. While most organizations see data as the source for gaining a potential competitive edge, the historical lack of focus on ensuring data is treated as an enterprise asset has led to widespread degradation of data quality. It will be highly beneficial to your organization, and your enterprise AI program, if you invested the time and resources in evaluating and/or refreshing your current enterprise data governance. If you currently lack a cogent enterprise data governance program, you do not need to build a "Cadillac." There are some basic tenets from data governance frameworks that should be implemented as a part of your enterprise AI program that focuses on the data being leveraged for AI. It is important for individuals to succinctly explain the value of data governance to the C-suite, but it is especially vital to be able to do so in the "Age of AI."

Data Governance programs aim to ensure that enterprises can fulfil the promise of that treasure trove of data that they contain. According to the National Association of States Chief Information Officers (NASCIO), Data Governance refers to the operating discipline for managing data and information as a key enterprise asset. This operating discipline includes organization, processes, and tools for establishing and exercising decision rights regarding valuation and management of data. Key aspects of data governance include decision making authority, compliance monitoring, policies and standards, data inventories, full lifecycle management, content management, records management, preservation, data quality, data classification, data security and access, data risk management, and data valuation (Sweden, 2009).

For most organizations, the challenge with being able to explain data governance, and the benefits of an effective data governance strategy are twofold: 1. The "corporate attention span" of senior leadership who do not see immediate return on investment from data governance programs, and 2. The maturity of organizational human capital to understand what can often be perceived as esoteric and nebulous concepts around data governance and data strategy. The most successful data governance frameworks, like with the most successful AI best practices, are those that are firmly built around people. Data governance programs are successful when they successfully influence organizational behavior, and inspire a change in how each employee in an organizational value chain sees and treats data. At the same time, a holistic data governance program is representative of every facet of such a program, from cataloging, taxonomy, and metadata management to discovery and delivery of insights (information) via curated data assets.

Illustrative Example of a Data Governance Framework:

It is worthwhile to revisit your enterprise data governance framework in the context of your enterprise AI program, should your organization be currently using one. If your organization does not have a formal data governance framework, you will need to shore up your data footprint for your enterprise AI program, by incorporating the best practices from a data governance framework. There are a wide range of data governance frameworks, of varying complexities, available online to choose from. Predicated on basic data governance frameworks, there are even data governance frameworks available that are bespoke to your industry, and usually incorporate data considerations that are unique to your industry. For instance, a data governance framework that is customized to the healthcare industry, might focus on "data privacy" and "data security" from a HIPAA (Health Insurance Portability and Accountability Act), SPII (Sensitive Personally Identifiable Information), or PHI (Protected Health Information) perspective.

Uber, operating in over 70 countries and tens of thousands of cities, uses an easy-to-understand data governance framework as depicted in Figure 54 (Soni, 2022).

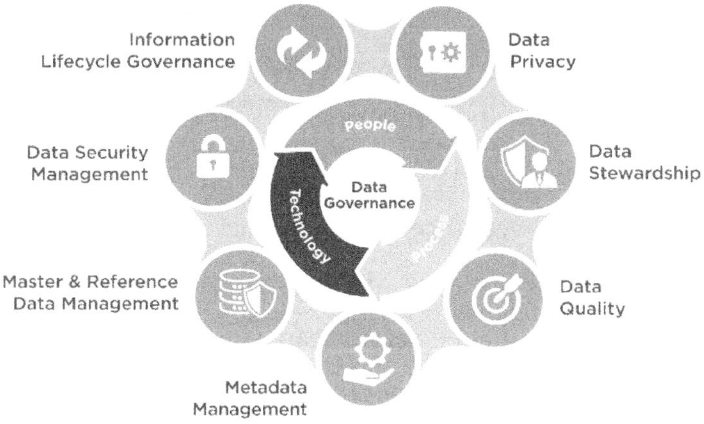

Figure 54: Data Governance Framework used by Uber (Illustrative Example) – credit in text

Use the Data Governance Framework used by Uber, any other appropriate framework of your choosing, or your own enterprise framework, as an illustrative example to explain/reiterate the importance of leveraging the tenets of data governance to business partners and other stakeholders. This would help ensure that your enterprise appropriately focuses on data in context of your AI program. The Data Governance Framework used by Uber is chosen as an illustrative example because like with the AIM Framework©, the central predicate of this framework is also the hallowed triad of People/Process/Technology.

The synchronized and coordinated partnership across a cross-section of stakeholders, distributed across multiple departments, is a critical facet of ensuring the initiation and ongoing success of any data governance program. Data governance, as aforementioned, is a cultural transformation for most established firms. Changing the culture of an organization requires people, the employees of your organization, to work in lockstep with each other and be aligned on the value of data governance and their role in it. Like with the AIM Framework© AI Governance Model, senior leadership support, and more importantly sustained senior leadership support is crucial. The lack of immediate tangible benefits resulting from data governance programs often make these programs vulnerable by being susceptible to being defunded. CEO support for this cultural change is vital.

The Uber example visually depicts the most critical functions of data governance required to engage with stakeholders and individuals unfamiliar with what data governance is, and allow to explain each component in the framework with specific examples in the value chain. It is noteworthy that unlike some other frameworks, this image clearly depicts an ongoing cycle. This is important since it is valuable to underscore and set the expectation that data governance is an ongoing program and not a defined project with a specific start and specific end.

Major Components of the Framework:

i. Data Privacy: With various regulations in place across states and countries, such as CASL, GDPR, PIPL, etc., knowledge of how to treat data, especially customer data that could contain Personally

Identifiable Information (PII), is of paramount importance. Data privacy governance in smaller organizations typically goes together with data classification rules.

ii. Data Stewardship: Another aspect of the framework that directly involves humans at the center of making data governance successful is stewardship. Data governance cannot be seen as the responsibility of an IT department or of a Chief Data Officer. Every organizational unit must play an equitable part in the care, management, and custody of their enterprise data assets.

iii. Data Quality: An aspect of data governance that is most well-known is one of data quality. The reason that data quality is most well-known and understood is because all employees of an organization, at some point, have been exposed to numerous examples, big or small, of data quality issues. Sometimes these are intercepted and fixed, but often, they are assumed as a cost of doing business and workarounds are established to step around challenges resulting from poor data quality. Most firms have some data quality automated checks in place, but most firms also have challenges with quality assurance and quality control with systems that contain legacy data.

iv. Metadata and Master Data Management: A renowned issue across enterprises is not that they have a lot of data, it is that they have a lot of uncategorized data everywhere. They lack a central taxonomy, and their data assets are referred to in a myriad of ways, diluting the enterprise's intellectual capital. Master and Metadata Management allows for providing a dictionary, taxonomy, and categorizing of data that facilities lucid discovery, reuse, and establishment of provenance.

v. Data Security: Data at rest and data in transit are critical elements of proper handling of data. In today's world where cybercrime is rampant, cybercriminals are essentially after data as a target, or seek to weaponize/hold hostage a firm's data.

vi. Information Lifecycle: Data to some - is information to others, and vice versa. As the definition goes, information is data imbued with value. However, information for one system or group can be consumed by another group as a data feed. Ergo, it is important to track not just data governance but the entirety of the information delivery value chain.

To Consider:

While the framework used by Uber is easy to explain to all stakeholders, it does have some shortcomings. I would personally prefer to highlight, in any framework, that people are at the core of data governance. Second, there are a few other important facets that I would consider as being absent from this framework. To overcome these shortcomings, note that in addition to the Sirius framework, consider leverage components from the Data Management Association (DAMA) as depicted in Figure 55 (Earley, Henderson, & Sebastian-Coleman, 2017). The DAMA framework also specifically includes "Data Warehousing and Business Intelligence" as well as "Data Architecture" as key components of data governance. To paint a holistic picture, the framework should start with Data Architecture and conclude the value chain with Business Intelligence and Analytics (systems of insights/systems of engagement).

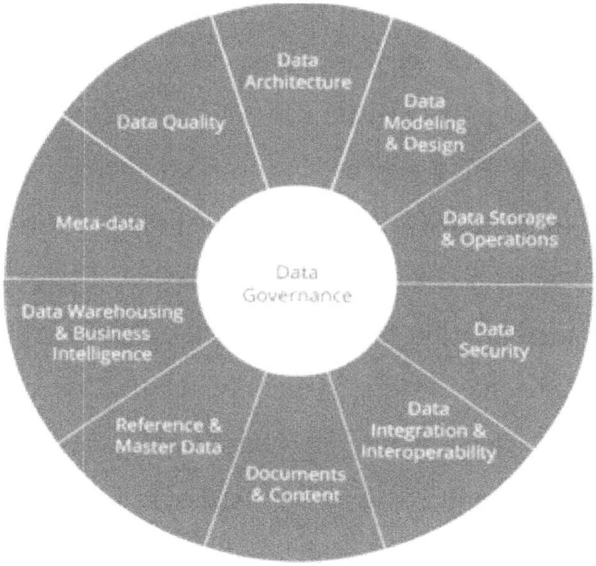

Figure 55: The DAMA Data Governance Framework

Real-World Data Challenges

The efficacy of your enterprise AI program is clearly dependent on the availability, efficacy, and quality of data obtained via external sources, as well as data residing within your ecosystem. It is likely easier to obtain curated datasets from external sources – data that you can have more confidence in – than it is to attest to the veracity and quality of your own data, thanks to the rigor of data providers, such as technology vendors.

Organizations often lack discipline, organizational muscle, infrastructure, and cannot rely on a bedrock of a data-driven culture, for being tenacious at executing their data and information management programs. Historical data management practices can fall short of delivering promised business value despite a significant amount of time, money, and resources having been invested into these data programs. Industries are seeking to adopt AI at scale, and continue capitalizing on their data assets by surfacing analytics and insights. While the C-suite is intently focused on transparent and explainable AI, there is an opportunity for business and technology practitioners to facilitate a candid discussion of the current state of their data practices. Established industries have built their data management solutions on a 'stack and silo' type legacy architecture within their systems footprint. As technology advanced, in order to kick the modernization can down the road, they have stacked platforms atop one another, and done so in their departmental silos.

Across industries, for companies with data strategies in place, the traditional methodology of how these companies have been seeking to manage their enterprise data - from discovery, to federation, surfacing, and consumption - needs to be revisited as a foundational need for scalable AI.

The pace of change in business challenges the ability for data management programs to deliver business value in a timely manner. While most data management practices, from warehousing to establishment of data marts and data lakes had been intended to federate disparate organizational data, the greatest challenge of bringing all the data in one place at one time, is the accelerating pace of business change. The larger the organization, the broader and deeper the dataset that needs to be coalesced to derive any meaningful insights, and deliver business value from the totality of an organization's data assets. Unfortunately, this also results in protracted cycles for development and delivery of these systems, by which time the business need evolves. It is often tenuous for business users to be able to prepare the data for consumption and conversion into meaningful and actionable information. Organizations, following a stack-and-silo based architecture have made it difficult to acquire the totality of a firm's full dataset. This makes the process of finding pertinent data challenging.

While storage and computational capabilities have gotten significantly better due to cloud-at-scale, the problem remains one of discovery, collection, transformation, taxonomy, and making curated data assets discoverable. The longer a data management program goes, from architecture to data acquisition, transformation, storage, discovery, and surfacing, it becomes harder for organizations to justify the return on investment (ROI) of these programs. Data management programs without being aligned to business objectives and drivers, can come across as nebulous, and esoteric. Absent clear and tangible outcomes, these programs can likely deliver on historical reporting, but lack the ability to provide any predictions. When enterprises that say that they want to be "data-driven," they usually mean that they are in pursuit of business predictability via predictive analytics. The paradox with these organizations is that they can become adept at historical reporting, but lack the ability to leverage their historical data to be able to achieve predictability. In that sense, these firms end up being able to "***predict the past***," when they should be looking to "***forecast the future***."

As a cumulative result, executives can get weary of funding multiyear programs that do not produce immediate tangible results. Contrast this with how executives view systems development today, with agile delivery allowing for stakeholders to visualize or interact with MVPs delivered expeditiously; MVPs that can commence providing business value.

Information Sharing

Information sharing across departments that traverse your enterprise value chain is essential for the success of any AI program. The surfacing, sharing, distribution and dissemination of information and information products across an enterprise, resolves and remediates several organizational problems and helps the firm to potentially capitalize on a plethora of nascent opportunities, while introducing a new set of challenges. The greater an organization's departments have been historically operated as silos, from a process and systems perspective, the broader the scope of challenges that are uncovered is likely to be. Sharing information across departments, and ergo improving the value derived from information products, provides for significant cost savings opportunities. These cost savings could result from either cost avoidance resulting from having the ability to render data-driven decisions, or because of operational efficiencies that are derived from being able to freely share information across business units. Often, the ability to share information allows a firm to reimagine or develop new business processes that have been institutionalized as workarounds. Workarounds are implemented to circumvent a poorly designed business process, a result of operating in a silo, or limitations within a system or technology product. Redundancy of systems that manage information products are also exposed and provide an opportunity to coalesce platforms and mitigate the application ecosystem "stack and silo effect" that has been mentioned earlier.

Sharing information allows for capitalizing on assets that had previously been under the custody and oversight of one business unit, such that redundant build, maintenance, and other investments can be eliminated. Information sharing allows for the organization holistically to gain visibility and insights into new revenue opportunities. These could come in the form of gaining additional insights into the purchasing preferences of their customer base, helping inform strategies on customer retention, new product ideas, and insights into new markets. When a company focuses on its customers, be it for attraction, or retention, they leverage customer relationship management systems to guide and inform their decision-making. This customer information, if relegated to a single department, disallows the firm from being able to understand their customer and inhibits their ability to provide an exceptional or customized customer experience. Information sharing can provide the company's senior leadership with a holistic and 360-degree perspective into the totality of the firm's operations, revenue opportunities, and risks, allowing for effective risk management or mitigation strategies. This perspective provides the decision makers with the ability to make data-driven decisions holistically and with increased confidence. As such, information sharing across an organization can provide a competitive advantage for a company over others.

However, organizations should also be watchful for inadvertent side effects that can manifest because of this information sharing. Improving information quality due to sharing information across the enterprise can surface an adverse impact on data quality within the organization. The primary driver for this is that with greater visibility and insight into enterprise information comes the challenge of the need to ensure efficacy of the underlying data, and this in turn highlights data quality issues that persist within departments. These challenges can be exacerbated for large, global, or multinational firms. Companies usually have a substantial challenge in trying to overcome their data challenges domestically. From legacy systems within established and mature companies, to technical debt, to the fact that data had historically been treated as a byproduct of technology systems, the challenges domestically are manifold. When amplified on a global scale, these data issues can be positively onerous.

During the process of information sharing and being made available across the enterprise, business processes that had been germane to a particular division or department are brought to the forefront. This also exposes inconsistencies in the underlying dataset that supports these business processes. Institutionalized workarounds specific to a department make it much more difficult to reconcile data assets across the enterprise. When companies seek to implement enterprise systems that are shared in use and value delivery across the company, such as CRM or ERP systems, especially in replacement of legacy department-specific solutions, data issues are brought to the fore. These could range from how departments use customer names, addresses, identify companies and customer IDs, leverage the free form "notes" fields, etc. Bringing these elements together to build a singular platform becomes difficult. This is made more challenging when there are mergers and acquisitions in an industry due to the disparity of cross-company systems. Another such challenge can present itself when companies undergo external audits, ranging from an audit of systems to SOX compliance, to attestation for compliance with data privacy and protection and data classification rules. Having cross-department information being made available surfaces potential violations or non-compliance of these regulations and guidelines within specific departments.

This chapter examined the importance of refreshing your data governance program when executing an enterprise AI strategy. Absent a data governance program, we explored a simple data governance framework that your organization could extract elements from, while establishing your enterprise AI strategy. This chapter also explored some real-world data problems that your organization can anticipate during implementation of an AI program. The next chapter concludes our journey through the topic of data in an AI context.

Chapter Twenty-Nine: Data, Data, Data – Part 2

We conclude our exploration of the importance of visiting or revisiting data governance in this chapter by examining data that resides within legacy systems, and the criticality of data literacy across your employee base.

Data in Legacy Systems

An established and mature organization, with several legacy systems across its functional domains, contains a vast amount of organizational intellectual capital within these systems. The data resident within these legacy systems is crucial for the organization to be able to undertake most strategic AI initiatives. These AI initiatives are predicated on the use of clean, accurate, secure, and reliable data. Consider a company that seeks to develop an outstanding and frictionless digital customer experience. If this company has the entirety of its customer information in a modern customer relationship management (CRM) platform, but a portion of customer data (such as billing history) resides within legacy systems, it is difficult to present an accurate, holistic view of each customer. Legacy system data is also notoriously difficult to work with and not easily queried.

A vestige of legacy systems and information architecture, data in these systems can potentially also be duplicated in multiple locations and across a multitude of databases. Typically, for a company with a monolithic legacy system such as a legacy CRM system, "the golden source of truth" can be found in their enterprise data warehouse. However, this golden source of truth is often actually replicated several times across an organization's ecosystem, devolving from a central system of record to multiple copies of the same. This makes it extremely challenging to identify and discern the true single source of truth, and renders it nearly impossible to develop a 360-degree customer profile. Organizations that set out to establish their "golden source of truth," can end up with multiple "golden sources of truth."

Legacy systems make it difficult to query data, with great effort required to match this data against data in other systems. It is challenging to get data into - and even more difficult to get data out of - these legacy system databases. The challenge that legacy systems pose is not only one of data identification and classification, but also inherently one of questionable data quality. Data quality evaluations for legacy systems are typically only surfaced and managed during modernization efforts. Without focused attention on the veracity and quality of data, bad data can foreseeably compromise the efficacy and integrity of AI models.

Data privacy, classification, and protection in legacy products are also notoriously difficult challenges. Some legacy products do not make it easy to protect or encrypt data-at-rest, leaving information security vulnerabilities across an enterprise. Without appropriate data classification, misuse of data can very easily occur. With numerous data privacy regulations on the books, multinational companies might find it increasingly onerous to comply with Data Privacy Regulations such as GDPR in Europe and U.S. state-based regulations.

In addition to AI implementations, operational reporting can also be sullied by bad legacy data. Most organizations that seek to be data-driven rely on reports that include data originating from these legacy systems. Without attention to modernizing these platforms (and, implicitly, evaluating the quality of the data residing within them), it is foreseeable that operational reports can be skewed. In a data-driven culture, the quality of these reports can mean the difference between making decisions based on sound data and potentially making decisions with incomplete or bad data.

Challenges are exacerbated when companies seek to modernize their legacy platforms, and when mergers and acquisitions happen across industries, where two or more organizations, with completely different systems of data management are expected to come together and integrate their data assets. Business users also spend a considerable amount of their time preparing data for consumption. All these challenges make it difficult for businesses to look upon data as a product, not merely as a by-product of their systems. Data governance, a practice that sets and manages the rules on how to treat organizational data assets, is more complicated in these types of fluid situations, with corporate structures and cultures rendering this problem worse. Without reimagining your established data management processes, you run the risk of deriving only a modicum of the benefits that data management, and therefore, AI, can offer.

A facet of the data that resides within legacy systems that your organization should be particularly sensitive to is that of inherent bias. Data resident within legacy systems is often a reflection of prevailing societal and cultural norms at the time of creation of that data. Societal norms have historically been biased. Data that is reflective of these societal norms can contain inherent bias, such as data that discriminates against a class of individuals. When utilized for the purposes of AI, without appropriate care and sanitization of this data, this inherent bias can sully even the most pristine AI models. Organizations that use AI models for the purpose of rendering decisions on humans, such as with loan processing, mortgage, insurance, etc., try to mitigate this issue by limiting the dataset that feeds their AI models to a certain number of years.

Often, companies seek to implement a new information system as a response to addressing information quality issues, or issues within legacy systems. In such scenarios, firms tend to focus on the new system and its implementation. This focus on the metaphorical tip of the iceberg skews the perspective of the inherent data challenges resident below the waterline. The reconciliation of disparate formats and data elements when sourcing to the new information system is laborious and resource intensive. Wrought with error and surfacing issues previously unknown, these projects tend to spiral in cost and time. Furthermore, aggregating poor quality data into a high-quality information system leads to a high-quality information system with poor quality data. This in turn leads to discontent with the implementation, resulting in mixed success for the initiative. Unable to see the return on their investments, organizations then shy away or defund data quality programs for a perceived lack of benefits. Defunded data quality initiatives are notoriously challenging to resuscitate since leadership, scarred once by their perception of failure, are reticent to allot funding to them in the future.

Below the Waterline of the AI Iceberg: Equipping Organizations for AI Success

The following is an article that I authored in the award-winning publication, LIMRA MarketFacts in June of 2023. Focused on the insurance industry, this article espouses the importance of focusing on data during AI program implementations in 2023. The central message of this article is pertinent here. Published with approval from LL Global, Inc. (Appendix B), this article, entitled “Beneath the Tip of the AI Iceberg” is as appeared in LIMRA MarketFacts, June 2023.

- START OF ARTICLE -

“Once a new technology rolls over you, if you're not part of the steamroller, you're part of the road,” according to Stewart Brand, American author and entrepreneur.

Artificial intelligence (AI) is not just the wave of the future; it is now, and every organization in every sector is positioning to capitalize on the opportunities.

There are innumerable benefits to deploying AI across the insurance value chain, including delivering a frictionless customer experience, understanding customers better, automating rote and repeatable tasks, introducing operational efficiencies, taking advantage of cost reductions, and that’s just the beginning. We haven’t even scratched the tip of the iceberg when it comes to the utility of ChatGPT and generative AI.

Industry leaders will be well-served to invest time and resources now in setting their organizations up for future AI success, which will be predicated on investments they make in their organization’s data assets.

Data Versus Information

Data has little meaning without context and purpose. Information is data endowed with purpose. AI, a broad term, alludes to machines — not humans — being able to consume vast amounts of data, and given context, do something purposeful with the data to generate information. However, any AI is only as good as the data it receives. Most mature industries, including insurance and financial services, are generators and consumers of vast amounts of data. However, paradoxically, we have underleveraged our organizational data assets, having treated data as a by-product of our systems and digital transformations, and not as a product.

In such a scenario, it is challenging to attain AI scale across the insurance value chain with data scattered across system silos of uncertain quality.

A good rule of thumb for industry leaders to espouse across their organizations is "Great AI + bad data = terrible AI."
The mindset around enterprise data will prove to be one of the key inhibitors of AI success within organizations.

Equipping AI Success

Companies should focus on three areas to ensure that data challenges do not deter the success of AI implementations:

Legacy Systems and Technical Debt

The amassment of enterprise legacy systems continues to be the albatross that stymies strategic investments. In addition to requiring diminishing and niche skillsets, being expensive, brittle and fragile to operate and posing cybersecurity challenges, legacy systems are notorious for poor data quality. This leads to a recurrent challenge within organizations; they transact business on antiquated systems with incomplete and poor data, albeit data that is necessary for the broader organizational AI goals.

Data Strategy and Governance

Organizations with a robust data strategy and governance program will reap positive results from their AI investments. Data management (data governance in practice) not only means having access to enterprise data in a central location, such as a data warehouse or data lake, it includes ensuring that data issues are at the forefront of a firm's objectives and that sound data management techniques are ensconced in daily operations.

Data-Focused Culture

Transparency and Ethics – Organizations will do well to focus on transparency and ethics. The explosive growth of AI and the "AI arms race" has led to calls to pause and establish ethical parameters and guardrails around the technology. No industry or sector has a comprehensive playbook for AI. The field is new, untested, fluid and changing so rapidly that developing such a playbook today ensures that it will be outdated tomorrow. The ethics of AI and around AI are a prominent concern across industries. The safe and effective use of AI by virtue of a framework known as explainable AI (XAI) seeks to solve the challenges with ungoverned AI.

These concerns include the lack of transparency for how an AI algorithm arrives at its decisions, and the sheer near impossibility that a human will be able to comprehend and retrace the decision-making pathways and processes that a machine — capable of millions of computations a second — makes. If humans are unable to understand how AI arrived at a conclusion or recommendation, it becomes difficult to trust the decision. Regulated industries, such as insurance, must ensure they are complying with emerging regulations to protect the consumer.

Data and AI as Strategic Enterprise Priorities – Firms that will lead the AI race will prominently feature AI and data in their corporate objectives, tie them directly to business goals and prioritize appropriately. Data-related programs across our industry that focus on quality and efficacy are especially susceptible to a lack of sustained focus and funding. The focus on data needs to be espoused to ensure that as the enterprise undertakes digital transformations it isn't simply focusing on new technology and systems, without solving for the data issues resident within the organization.

C-suite Champions – Once AI enterprise-level objectives have been established, it will be vital for the C-suite, especially the CEO on down, to stress the importance of data across the organization.

Structural Support – The chief data officer or chief data analytics officer is a relatively new role across the insurance industry. Firms would benefit by augmenting the role to include and reflect AI responsibilities. Depending on their maturity, firms should consider creating a chief AI officer role.

Enterprise Data Literacy – MIT defines data literacy as "the ability to read, work with, analyze and argue with data." According to a Gartner study, only one-third of employees across an average organization can confidently understand, analyze and argue with data. This implies that two-thirds of people in most firms might not be able to consistently discern good data from bad data.

Bad data costs the U.S. $3 trillion annually. Studies show that 40 percent of enterprise data is either inaccurate, incomplete or unavailable, which results in businesses failing to achieve data-driven goals; the cost of bad data is between 15 to 25 percent of revenue for most companies. It is imperative to educate employees about the value of data and measurably increase data literacy, empowering employees to solve data issues as they encounter them.

The Road Ahead

To mitigate risk and capitalize on opportunities, business and technology leaders must develop and bolster the structural integrity of AI programs; in addition, they must create a data-driven culture. Making these investments now just might be the deciding factor between riding the crest of the coming AI tidal wave or being deluged by it.

- END OF ARTICLE –

Special thanks to LIMRA and LOMA (LL Global, Inc.) for their permission to reprint this LIMRA MarketFacts article.

Data Literacy

The concept of Data Literacy is broached in the LIMRA MarketFacts article above. Data Literacy is of great importance to a successful data management program that is founded on sound data quality. Implicitly, accurate and reliable AI that is driven by accurate and reliable data, is driven by employees across your value chain that can discern the difference between good data and bad data. Unless employees understand – and are literate – about data, they might be unable to tell good data from bad data. These are the individuals that organizations are most reliant on to serve as custodians and purveyors of data. They are the most hands-on with data on a daily basis. If employees are uneducated or undereducated about the value of sound data quality, and they cannot discern between good and bad data, they might inadvertently negatively impact the efficacy and quality of your AI strategy by sheer inaction alone.

Definitions

There are several definitions of Data Literacy that are largely like each other. Below are three of the more popular definitions of Data Literacy, with my own derivation as the fourth.

1. ***Gartner***: The ability to read, write and communicate data in context, including an understanding of data sources and constructs, analytical methods and techniques applied, and the ability to describe the use case, application and resulting value (Panetta, 2021).

2. ***Massachusetts Institute of Technology (MIT)***: Data literacy describes the ability to read, work with, analyze, and argue with data (Brown S. , 2021).

3. ***Qlik***: Data literacy is the ability to read, work with, analyze and communicate with data. It's a skill that empowers all levels of workers to ask the right questions of data and machines, build knowledge, make decisions, and communicate meaning to others (Qlik, n.d.).

4. ***My Definition***: Data literacy is the ability of your employees to be able to read, analyze, synthesize, share, and leverage your high-quality data assets using a shared understanding, in order to help your customers with data-driven confidence, increase internal efficiencies, and optimize your revenue potential by allowing you to monetize and apply your data.

Figure 56 depicts five groups of questions that organizations should use for self-reflection when considering improving your enterprise Data Literacy.

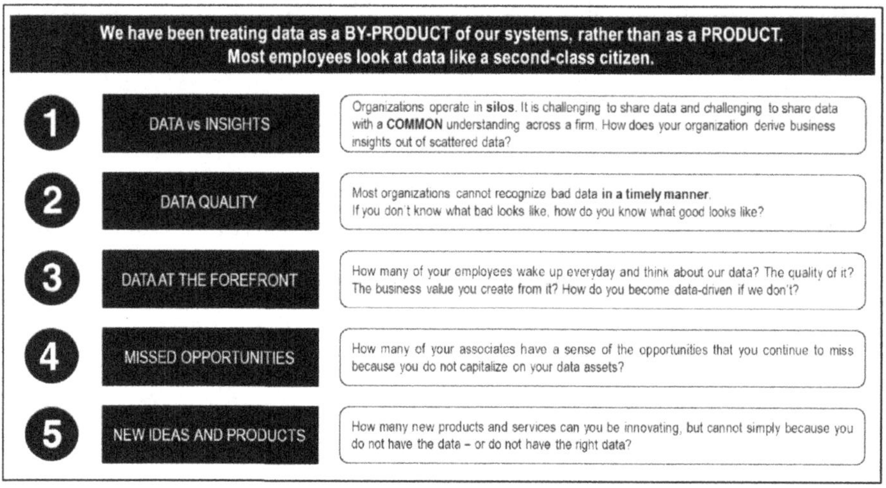

Figure 56 – Self-reflection questions when considering Data Literacy

Data Literacy Challenges

The most common challenge with successfully nurturing a data-driven culture that nurtures an employee-base that is literate in data, often comes down to lack of clarity around roles and responsibilities – specifically ownership of enterprise data. Data is a business issue that has long masqueraded as an IT issue. Data can serve as one of the primary sources of tension between the Business and IT. This tension stems from the question of who is primarily responsible for data quality. The Business perceives IT as the arbiters of data quality since IT is responsible for managing data and the multitude of systems that produce, deliver, and consume data. IT sees itself as the builders and support of systems, and perceives the business to be responsible for the data, and implicitly, the quality of the data that flows through these systems. Figure 57 depicts the ownership of data within an organization as a reminder for your employees on who data belongs to. Each of the five statements below are true. This lack of clarity can be antithetical to your enterprise AI strategy by sowing confusion about ownership of enterprise data.

✓	**Data does NOT belong to IT**
✓	**Data does NOT belong to the Business**
✓	**Data belongs to IT**
✓	**Data belongs to the Business**
✓	**Data is EVERYONE'S Responsibility**

Figure 57: Ownership of data within an enterprise

There are very real costs associated with poor Data Literacy. Figure 58 depicts the challenges because of poor Data Literacy.

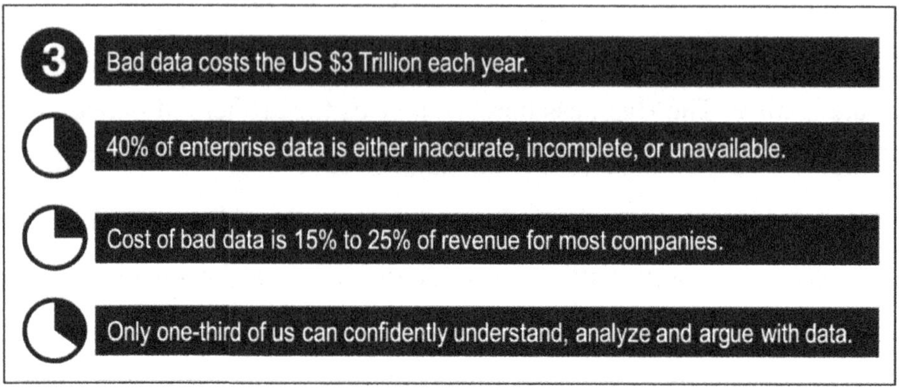

Figure 58: Impact of poor Data Literacy by the numbers

Figure 59 depicts the 1:10:100 rule. The 1:10:100 rule shows the costs associated with resolving issues due to bad data. According to the 1:10:100 rule, it costs $1 to fix bad data at the point of creation (Prevention Cost). The same bad data, if it enters a development/test environment during the software development cycle, costs $10 to remediate. This is known as the Cost of Correction, or Correction Cost. If the same bad data goes undetected and enters a production environment (live), and is reported as a defect when operational, costs $100 to fix. This is known as the Failure Cost.

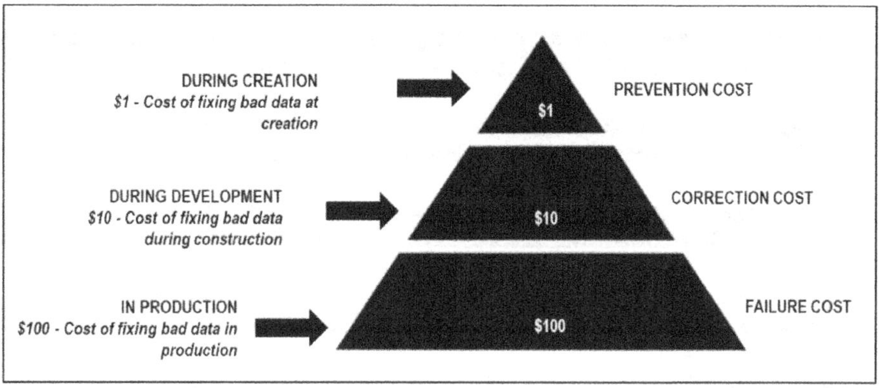

Figure 59: The 1:10:100 Rule of Bad Data

This is why Data Literacy is so important. If an employee is not data literate, they will be unable to intercept and remediate bad data at the source, or at least, during the development cycle. Absent an employee-led focus on data quality, your good data will always be sullied by bad data. It will be challenging to determine predictive value of data and get insights, and you are likely to miss better data insights in an ocean of everyday data insights.

Good data is core to an efficient, optimized, thriving business, and data literacy is core to be able to derive any value from your data assets. Your enterprise AI program is reliant on an employee-base that is literate - if not fluent - in aspects of data that is pertinent to their jobs. Figure 60 depicts the top ten challenges your organization will face absent a focus on Data Literacy.

1	**Opportunities Abound**	Unable to anticipate and capitalize on new / expanding market opportunities, internal efficiencies.
2	**Stymied Growth**	Limited growth potential, competitive pressures.
3	**Shoehorned Technology Ecosystem**	Technology ecosystem cobbled together with workarounds for matching data.
4	**Hardening Enterprise Silos**	Departments entrenched in their own systems, own processes, and their own data .
5	**Anchored Down**	Data - especially from legacy systems – will weigh down the organization.
6	**Significant Expense**	Good data practices can make you money, but bad data practices **WILL** cost you money.
7	**Analysis Paralysis**	Employees spend significant time poring over all kinds of data to make decisions.
8	**Tech Cart Before The Business Horse**	Companies lead with systems and technology instead of data-driven business decisions.
9	**Potential Risks Below The Surface**	Unaware of data risks – classification, security, privacy and protection.
10	**Hatching The Next Big Idea**	Limited ability to develop new products and services.

Figure 60: Ten Issues Resulting from Poor Data Literacy

Five Steps to Data Literacy

Achieving Data Literacy across your organization is not a one-and-done event. Educating employees on the value of being data literate, and investing in their training and development, will have positive compounding effects across your enterprise well beyond your enterprise AI program. Data Literacy is at the foundation of your data governance and data management program, which in turn will provide a strong base to your enterprise AI strategy. As Figure 61 depicts, Data Literacy is at the base of the Information Iceberg, with the real work of focusing on your enterprise data residing below the waterline of this iceberg.

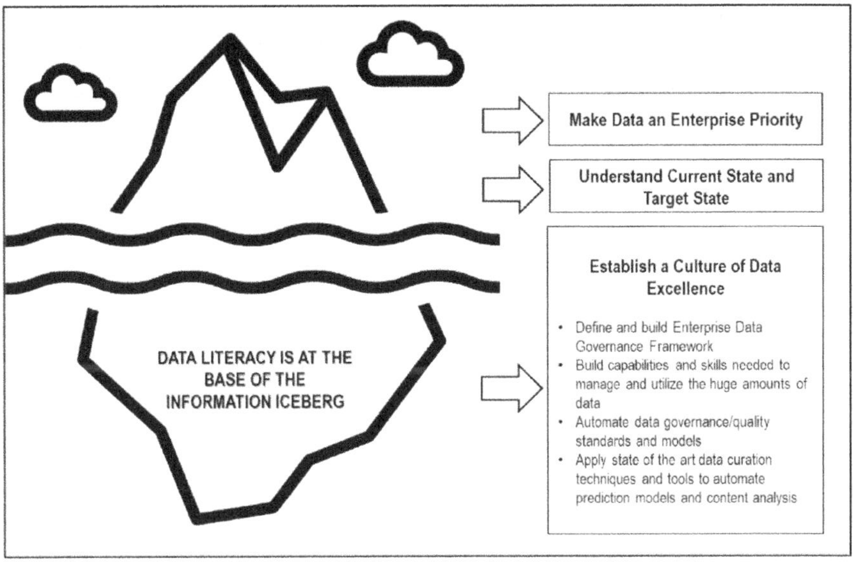

Figure 61: Data Literacy at the base of the "Information Iceberg"

Here, at a high-level, is a five-step process to achieve an ongoing focus on Data Literacy:

1. ***Make Data an Enterprise Priority***: Incorporate and feature data as a priority in the corporate vision and mission.

2. ***Communicate from The CEO On Down***: Your CEO HAS to set the tone. This tone NEEDS to be amplified by every leader across the enterprise.

3. ***Think Global, Act Local***: Set ENTERPRISE standards for people, process, technology, but empower LOCAL implementation.

4. ***Train and Educate Employees***: Formalize industry education AND data education as part of your learning and development plans.

5. ***Engage and Empower Employees***: Communication, repeat, reiterate – BUILD a CULTURE around data.

SECTION THREE

AI BEST PRACTICES

PART E – LEADING AN AI-READY ORGANIZATION

Chapter Thirty – Leading in the Age of AI

Visionary and steady leadership is a vital part of any organization's success. Leadership is even more critical during times of accelerated change, and to navigate through the resulting uncertainty due to this change. The "Age of AI" will place a large spotlight on leaders across organizations as they steward their companies through unprecedented and accelerating change. The criticality of steady leadership through the "Age of AI" cannot be understated. It will serve as a guiding force for your organization, shaping the future, vibrancy, the evolution, and even the potential survival of your company over the next several decades.

Navigating your organization through the evolution of AI through the first half of the 21st century is akin to flying a plane full of passengers, through the Alps, at low altitude, while wearing a blindfold. It will require adept, skilled, calm, and stoic leadership to be able to successfully do so. The role of a leader and the impacts of their decisions are more prominent now than ever before. Those leaders that are able to grasp the complexities of AI, capitalize on the transformative nature of the technology, and guide their organizations with a clear vision and strategy, nurturing a fail-resistant corporate culture that thrive amidst change, will be the ones who will architect their companies to attain sustainable success with AI.

The AIM Framework© has underscored the critical importance of the CEO and the C-suite setting the tone for AI within your organization. Readying your firm for the "Age of AI" requires visionary leadership, strategic planning and execution, and a commitment to fostering a culture of experimentation and continual learning. However, leadership across all facets of your enterprise AI strategy is not limited to the CEO and the C-suite. When it comes to leading your organization through AI-driven changes and executing on your enterprise AI program, every leader across the enterprise has to be aligned with strategic execution, and serve as a force multiplier. Starting from the top-level leadership, through to the middle managers, to frontline team leaders, the successful execution of your enterprise AI program is entirely predicated on the ability of leadership to successfully effectuate a cultural transformation of your enterprise, as much as it is about the underlying technology and supporting processes. A people-centric approach towards AI leadership is fundamentally vital for sustained AI success. This people-centric approach requires leaders to shepherd their organizations, capitalizing on the transformational power of AI to enable existing business goals, expose new business opportunities, create value, establish their enterprises for long-term, responsible AI success, develop and augment the skills of their employees, and champion experimentation and innovation.

Leading an AI-Ready Organization

Leadership in the "Age of AI" is not a job, or a position – good leadership will help shape the story of your sustained success with AI. Being exemplary leaders in the "Age of AI" demands a different approach. In addition to being champions and stewards of the Basic Principles and Enterprise Best Practices outlined by the AIM Framework©, leaders should expect to intently focus on change management. There are a few key aspects of a leader's responsibilities and focus areas that are outlined below.

1. Vision and Strategy:

As Enterprise Best Practice 1 has outlined, outlining a vision, and developing and communicating an enterprise strategy around AI, is the primary responsibility of the organization's topmost leadership. Commencing with the CEO and the C-suite, the establishment of a clear vision and AI strategy is a predicate to your enterprise AI program. Leaders must ensure that their vision and strategy around AI are not separate and distinct from their business goals and objectives.

Whether it is to capitalize on new opportunities, strengthen existing strategic objectives, or to mitigate new and existing risks, leaders must incorporate AI as enablers to business goals.

Continuously monitor and evaluate the impact of AI initiatives on the organization. Establish metrics and key performance indicators (KPIs) to track the effectiveness of AI adoption. Regularly review AI projects, assessing their alignment with organizational goals and the value they bring. Collect feedback from employees, customers, and other stakeholders to identify areas for improvement and refine AI strategies.

Figure 62 encompasses five key considerations when building and communicating an enterprise AI vision and strategy.

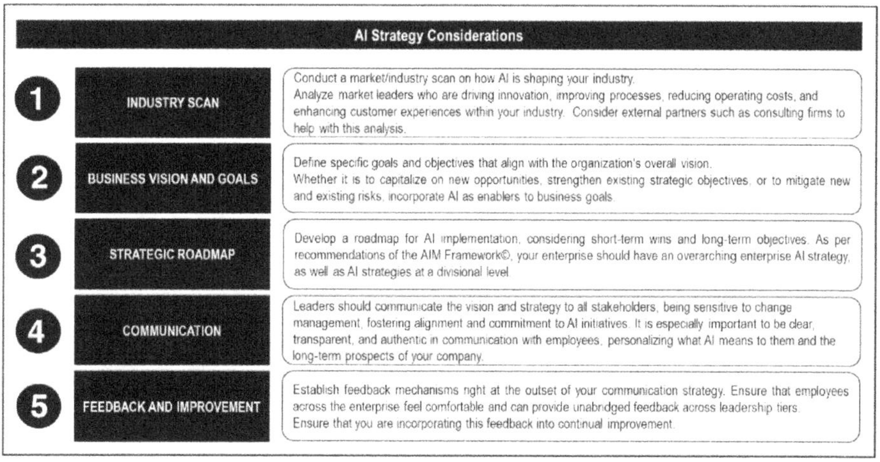

Figure 62: Considerations when building an AI Strategy

2. Employee AI Education:

A workforce that is educated on AI is one that will propel your enterprise AI strategy to success, whereas an undereducated workforce might jeopardize it. AI is a fluid, vibrant, and rapidly evolving field. Your employees would immensely benefit from being exposed to the fundamentals of AI, and its safe and effective use within your organization, with targeted education of potentially increasing complexity based on role. Education, as a strategic focus, would lead to continual re/education and serve as an ongoing engagement vehicle across your employee base. A workforce that is educated on the fundamentals of AI, and how they can apply AI to their own jobs, can help your firm realize process efficiencies, automation, cost savings, etc. from an invaluable bottom-up perspective. AI education will ensure that your employees are diligent and aware of the safe and effective use of AI. Employees that are engaged in the AI dialog at an enterprise level will ensure that AI is not seen as just another technology initiative, nor will it operate within department silos.

Your enterprise AI education should serve to provide a level-set to employees on the fundamentals of AI. Your education program can then offer online learning trails based on existing content on AI aligned to potential interests and technical competencies. As your enterprise AI strategy matures, it will be imperative for you to develop learning and development, re/education plans for your employees. This education program should also create feedback mechanisms for employees to ask clarifying questions. It is recommended that your employee education program capitalizes on existing knowledge management solutions that are employed within your organization. This could include the publication of an internal forum for AI-related content and ongoing engagement vehicles. Consider publicizing external training opportunities, forums, events, etc. for employees to gain further/additional knowledge about AI, and make these a part of their career development plans/annual goals and objectives. When developing objectives for your employees that include education, you should consider changes to jobs over a five-to-ten-year span, and identify opportunities for retraining of your employees, commensurate to these changes.

Leaders must educate themselves on AI to understand the potential of AI and their impact on your industry and organization. Leaders should familiarize themselves with AI concepts – such as those in Section One of this book, and be aware of applications of AI that are relevant to your industry. It will be imperative for leaders to stay up-to-date and informed about the latest AI advancements through a variety of forums such as conferences, trade associations, and engagement with external consulting partners. This knowledge will enable leaders to make informed decisions, adjust AI strategies as required, and effectively manage the AI change curve within your firms.

3. Employee Enablement, Ethical AI, and Governance:

As detailed in the chapter on Explainable AI (XAI), leaders must prioritize ethical AI practices in order to build trust and ensure responsible AI implementation. Leaders are custodians of ensuring that the enterprise follows the highest ethical standards set by the firm, in addition to ensuring compliance with regulatory requirements. Leaders should establish guidelines and frameworks for ethical AI development at the outset. Your enterprise AI vision and strategy should be predicated on an immutable foundation of ethics and transparency. It should ensure that you are proactively addressing issues such as bias, privacy, transparency, and accountability.

Leaders must ensure that employees are aware of responsible and ethical use of AI, and can safely leverage AI to assist with their job functions and innovate with appropriate safeguards in place (contingent to your enterprise stance on granting employees access to AI tools). The CEO and C-suite must ensure that they involve cross-functional teams, including legal, compliance, and data privacy experts, in shaping your organization's ethical AI policies. Given the fluidity of the field of AI, leaders should ensure that they regularly review and update their AI guidelines as AI and regulation around AI continues to evolve. Enabling AI via an evolving enterprise AI Governance policy and enforcement mechanism is essential and serves as a crucial framework to ensure ethical, legal, and responsible AI use. This will not only mitigate potential risks and liabilities, but also help build trust with stakeholders, customers, and partners.

By establishing clear guidelines on safe and authorized AI use, bias mitigation, transparency, and accountability, your organization can minimize the chances of unintended consequences, or unethical AI behavior. Effective enforcement mechanisms help monitor responsible AI use, ensure compliance, and adapt to evolving regulatory landscapes, safeguarding your reputation and long-term success while fostering innovation. Leaders must ensure that their employees understand the policy, oversee the communication, rollout, and education of employees on AI Governance policy, providing a means for employees to ask questions and solicit additional information. Leaders should facilitate the incorporation of pertinent employee feedback into future revisions of the AI Governance policy. A workforce that is educated on risks is one that is actively involved in risk mitigation by protecting your Intellectual Property and safeguarding your digital assets.

4. Skilling and Reskilling the Workforce:

Preparing your workforce for the "Age of AI" is vital to the long-term success of your enterprise AI strategy. In addition to the initial educational efforts, leaders should help employee's skill/reskill/upskill themselves, and ensure that these skill development opportunities are incorporated into employee training and development plans. To successfully be able to help employees gain new skills, leaders need to be able to forecast jobs of the future, or how existing jobs will change over time, due to AI. Once your enterprise has been able to identify skills required in the future, leaders can conduct a current state assessment and identify skill gaps. Please note that this will be an imprecise science at best. It is going to be challenging to precisely identify jobs of the future, and job changes to existing roles. It bears repeating, a good rule of thumb would be to identify jobs that have rote, manual, repeatable tasks that can be automated by virtue of AI. Just as we could not have predicted in 2007, when the iPhone was first introduced, that the game "Angry Birds" would become a $1B franchise in 15 years, we will not be able to accurately predict what new industries and jobs AI will create (and disrupt). Leaders should ensure that their organizations invest in training programs, workshops, and partnerships with trade associations and/or educational institutions to upskill employees. Leaders at all levels must encourage a continual learning mindset, and be intentional about creating opportunities for employees to apply their AI skills in their jobs.

Planning for Reskilling

The rise of AI is a seminal moment that has the potential of automating over half of some roles. This will make it imperative for your enterprise to identify jobs that need reskilling or transformation. By recognizing these roles, you can strategically allocate resources to ensure your workforce remains relevant in the AI era.

By adopting data-driven approaches, staying informed about AI advancements, and engaging employees, you can ensure that your workforce remains competitive and adaptable in order to continue delivering value to your customers and stakeholders. Reskilling and transforming jobs affected by AI is an investment in the long-term success and sustainability of your company. It is a proactive step toward harnessing the potential of AI while preserving the value of human expertise. This is a strawman approach of key considerations and methods for identifying jobs that require reskilling or transformation due to AI. Leaders should consider conducting this analysis as an HR-led initiative that will require active participation across every division, led confidentially by the C-suite member/department head. Figure 63 presents some considerations as leaders plan for reskilling.

Adapting to AI Disruption: AI technologies are automating tasks across various industries, necessitating reskilling efforts to remain adaptable.	**Customized Learning Paths:** You will need to develop personalized reskilling programs to cater to individual skill gaps and interests.
Preserving Human Relevance: Reskilling empowers employees to focus on tasks that require uniquely human skills, such as creativity, empathy, and critical thinking.	**Cross-Functional Training:** You should consider encouraging employees to learn skills beyond their immediate job roles, promoting versatility.
Future-Proofing: Preparing the workforce for AI advances will help your firm stay agile in the face of rapid technological change.	**Leverage AI for Training:** You should plan to utilize AI-driven learning platforms to deliver adaptive and relevant training content.
Meeting Skill Gaps: Identifying the skills gap in the enterprise will be essential for tailoring reskilling programs effectively. This effort should include HR and divisional leaders.	**Continuous Learning Culture:** Through this reskilling initiative, you must foster a culture of continuous learning where reskilling becomes an ongoing process rather than a one-time effort.
Data-Driven Approach: Employ data analytics to assess which jobs are most at risk of automation and prioritize reskilling efforts accordingly.	**Certifications and Recognition:** You should plan to establish clear certification programs to validate new skills and motivate employees.
Engaging Employees: You should involve employees in the reskilling process, ensuring their input and needs are considered. Programs should further their career interests *and* business need.	**Mentorship and Coaching:** You should plan to pair experienced employees with those undergoing reskilling to provide guidance and support.

Figure 63: Planning for Reskilling

Driving AI-Driven Reskilling

Reskilling will emerge as a crucial strategy to ensure that your workforce remains relevant and competitive. The integration of AI presents both challenges and opportunities. To address this, leaders must identify which jobs require reskilling and develop effective strategies for upskilling your employees. Reskilling is not just an option, but will become a necessity in the "Age of AI." Leaders must proactively identify jobs for reskilling and implement strategic initiatives to thrive in this era, and also ensure your employees remain an asset in an AI-driven future. Reskilling is the bridge that connects human potential with the limitless possibilities of AI. Figure 64 provides an overview of how leaders can go about driving AI reskilling across your enterprise.

Data-Driven Analysis: Leaders should leverage data analytics to assess which job roles are most susceptible to automation or augmentation by AI. This analysis can include factors such as the nature of tasks, routine vs. non-routine work, and the potential for AI to improve efficiency.

Impact Assessment: Leaders should consider the impact of AI on job roles in terms of cost reduction, productivity enhancement, and quality improvement. Jobs that are most affected by AI's capabilities are prime candidates for reskilling or transformation.

Technological Advancements: Leaders must stay informed about the latest AI developments in our industry. As AI evolves, new opportunities and challenges will arise, necessitating a proactive approach to identifying jobs that need adaptation.

Employee Feedback: Leaders should engage with employees to gather insights about their daily tasks and the challenges they face. Frontline employees often have valuable perspectives on areas where AI could enhance or disrupt their work.

Skill Mapping: Leaders will need to map the skills required for each job within the enterprise and then compare these skill sets to the capabilities of AI. Jobs with significant overlap may need reskilling efforts to incorporate complementary skills.

Industry Benchmarks: Leaders should consider conducting a benchmark against industry peers to understand how they are adopting AI, and then identify areas where they lag behind or excel in AI adoption and adjust their reskilling strategy accordingly.

Job Redesign: Leaders will need to consider how AI can be used to redesign jobs. For example, certain routine tasks can be automated, freeing up employees to focus on higher-value add activities. This may require creating hybrid roles that combine human expertise with AI support.

Cross-Functional Training: Leaders will need to encourage cross-functional training to make employees more versatile. Jobs that require a broader skill set are more likely to adapt to AI advancements.

Pilot Programs: Leaders should consider implementing pilot programs to test AI integration in specific job roles, measuring the impact on productivity and quality to determine the feasibility of wider adoption or reskilling.

Industry and Customer Insights: Leaders will need to understand the changing demands of their industry and customer needs. Jobs that directly influence customer experience and engagement may require reskilling to meet their evolving expectations.

Figure 64: Driving AI-Driven Reskilling

5. Effectuating a Cultural Change:

Regardless of field, sector, industry, or any given part of the overall value chain, organizations will continue to face an AI-driven seismic shift that demands a fundamental change in the way they operate. AI continues to rapidly reshape industries, processes, and job functions, a transformation that will only continue to exponentially increase as AI continues to increase in sophistication and maturity. AI is a continuation of the transformative, and often disruptive, Digital Transformations that have been ongoing across sectors and industries through the first quarter of the 21st century. The distinguishing characteristic between Digital Transformations, and AI as a continuation of this journey, is that AI is unconstrained by the limits of human intelligence, and unencumbered by the intrinsically human feelings towards the pace of change and transformation. The commonality of the impacts of these Digital Transformations and AI within organizations is that both are fundamentally predicated on humans, and more specifically the human capacity for change. Within organizations, this human capacity for change – to be able to absorb and adapt to change instead of resisting it – is a core defining trait of an organization's corporate culture. As Peter Drucker has stated, "culture eats strategy for breakfast." As with any Digital Transformations, to thrive in this new era, organizations must prioritize change management and transformational leadership. This chapter delves into why this is essential for organizational success and outlines strategies for leaders to foster a climate where employees can embrace change and adapt to the AI revolution. The "Age of AI" requires organizations to be more agile and adaptive than ever before, and it is up to leaders to champion this adaptability and enterprise agility.

A culture of safe innovation is essential for successful AI implementation. It will be a leader's responsibility to nurture this culture of innovation by creating a fail-resistant organization, encouraging curiosity, experimentation, and risk-taking within the enterprise. Leaders should coach employees that they should consider failure as a learning opportunity. This includes ensuring that leaders are not quick to judgment and there are appropriately no punitive impacts of failure (as appropriate – of course there should be appropriate repercussions if uncontrolled experimentation accidentally results in significant damages). Leaders need to create an environment where employees feel empowered to explore leveraging AI. For AI to succeed, it will be vital to ensure that it does not operate in a divisional silo. Leaders must foster cross-functional collaboration to leverage diverse perspectives and expertise across the enterprise value chain. Rewards and recognition will be important for leaders not to lose sight of, and they should appropriately recognize and reward innovative thinking as positive reinforcement. In addition to facilitating collaboration across an enterprise value chain, leaders should foster collaboration between employees and AI to maximize their combined potential. Employees should be encouraged to think of AI as a tool that enhances their capabilities rather than a replacement and a threat to their jobs. Leaders must foster an environment where employees feel comfortable working alongside AI, leveraging AI-driven output to inform data-driven decision-making.

Most of all, preparing an organization for AI requires effective change management techniques. Be it a derivation of the ADKAR framework, or a change management framework that your enterprise currently uses, leaders should ensure that change management encompasses communication, training, and stakeholder engagement. It is imperative to communicate the rationale clearly behind AI adoption and transparently, addressing potential concerns in a proactive manner, and dispelling misconceptions.

SECTION THREE

AI BEST PRACTICES

PART F – "HUM-AI-N" PERSONAL AI READINESS

Chapter Thirty-One – "Hum-AI-n" – Preparing Yourself for the AI Age – Part 1

It sounds like a paradox, but **to succeed with AI**, to thrive amidst all this technological change, you must **be more human**. This is the central predicate behind the term "hum-AI-n."

The skills required to prepare yourself in the "Age of AI" are a microcosm of the skills required for preparing yourself for rapid technological change. The next several decades will witness an unprecedented convergence of technological advancements, with AI leading the charge. The opportunities for you to thrive in this era - professionally and personally - are vast, but they come with the responsibility to adapt, learn, and evolve.

AI-Driven Career Changes

The rapid evolution of AI will lead to the emergence of new industries and job opportunities. While some traditional roles may be automated, this technological shift will create a demand for human expertise in areas such as AI development, data analysis, cybersecurity, and human-AI interaction. Additionally, as AI takes over mundane tasks, there will be increased opportunities to pursue creative endeavors, entrepreneurship, and personalized skill development.

The rise of AI-driven change undoubtedly presents career risks, but those who proactively adapt can find success and fulfillment in the face of these challenges. By cultivating a growth mindset, embracing lifelong learning, fostering adaptability, leveraging transferrable skills, staying attuned to emerging opportunities, and fostering a supportive network, you can navigate the shifting career landscape and carve out a resilient and prosperous professional journey. The rapid pace of technological progress and automation will likely result in the obsolescence of many jobs that exist today over the next 25 to 50 years. Jobs that require routine, repetitive tasks are particularly vulnerable to being automated, and will likely be replaced by AI (and/or machines) that can perform the same work more efficiently and accurately. Some of the most vulnerable industries include manual labor, routine-based jobs, and certain administrative and office support roles.

With the continued development of automation and robotics, jobs that involve repetitive manual tasks such as assembly line work, packaging and labeling, and data entry are likely to become obsolete. Jobs that involve routine-based tasks such as bookkeeping, data analysis, and customer service are also at risk. The professions in imminent risk of extinction include white collar and blue-collar jobs alike. White collar - offices jobs - such as that of a data entry clerk will be extinct as advanced AI algorithms will automate data processing and analysis.

Blue collar jobs in the manufacturing sector and others that involve repetitive manual labor that would likely be replaced by robots and other AI. Consider that robots have been deployed in Japan for elder care and to address the needs of that country's aging population for several years through the first quarter of the 21st century. Robots have been in use to deliver food to college students across university campuses in the United States as of the 2020s. Robots serve as food delivery, effectively being waiters at restaurants such as Buffalo Wild Wings in the United States. Intentionally friendly-looking robots operate at Dubai International Airport for screening passengers for symptoms and markers of COVID-19. There are AI-powered robots that perform janitorial and custodial work. Robots, once considered a novelty less than five years ago, will with certainty become an integral part of industries and our daily lives globally over the next decade. The widespread use of drones is transforming everything from Amazon package deliveries and food deliveries using popular delivery services such as UberEATS and Grubhub to how wars are conducted.

AI is on the precipice of a milestone, where Artificial General Intelligence (AGI) will find itself dispersed throughout our everyday technologies such as with OpenAI's ChatGPT imminently finding itself being embedded across Microsoft Office products (Metz & Weise, Microsoft Bets Big on the Creator of ChatGPT in Race to Dominate A.I., 2023), and competing against Google as a viable search engine. Think about the traditionally white-collar professions that this will transform – from being able to automate the authoring of legal briefs to generating entire books – AGI will create as many more abundant new occupations as it disintermediates.

The stakes are extraordinarily high for those serving in professions that are at imminent risk of being made obsolete. To prepare, individuals in jobs that are at risk of being made obsolete need to invest in their education and skill development. This can be done through formal education, such as taking courses and obtaining degrees in fields that are less likely to be automated, or through on-the-job training and experience. Individuals must be proactive and seek out new skills and knowledge that will be in high demand in the future. This may involve upskilling or reskilling in areas such as data analysis, programming, and digital marketing. It's also important for individuals to stay up to date with the latest technological advancements and to be proactive about acquiring new skills and knowledge that will make them more valuable to employers in the future.

It is also important for individuals to stay informed about the latest AI advancements and understand the potential implications for their current or future careers. This can involve attending conferences and workshops, taking online courses, and networking with experts in the field. Another way for individuals to prepare for the impact of technology on their jobs is to develop transferable skills, such as critical thinking, problem solving, and creativity, which are less likely to be automated. It will be even more important in developing expertise in areas such as ethics, creativity, and emotional intelligence, which are unlikely to be easily and imminently replaced by AI. These skills are in high demand in a rapidly changing job market, and will make individuals more adaptable and competitive in the future.

As much as we need to prepare ourselves for the "Age of AI," it will also be important for individuals and society as a whole to consider the broader implications of technological unemployment, and to consider how best to support those who are displaced by automation. Governments and organizations, through us as individuals, must also play a role in helping people prepare for the future of work. This can include offering job training and education programs, unemployment insurance, government support for retraining and skill development, creating incentives for businesses to invest in upskilling their workforce, and developing policies that support workers who are displaced by AI-driven change. While the automation of many jobs will likely result in significant disruptions to the labor market, we can prepare for this change by investing in their education and skill development, acquiring transferable skills, and supporting initiatives that help to mitigate the negative impact of technological unemployment. By being proactive and seeking out new skills and knowledge, we can position ourselves for success in the rapidly changing world of work.

Leveraging AI as a Tool

As AI continues to evolve, traditional educational models need to evolve to keep pace with the evolving demands of the job market. You will need to continuously update your knowledge and acquire new skills to remain relevant. Adapting to new technologies, such as AI systems, robotics, and automation, will require you to develop a foundation in STEM (Science, Technology, Engineering, and Mathematics) fields while also nurturing creativity and critical thinking abilities. Corporations, governments, educational institutions, and we, as individuals, must invest in reskilling and upskilling programs to equip the workforce of tomorrow with the necessary competencies required in a technology-driven society. Embracing a growth mindset and cultivating curiosity will enable you to embrace change and seize new opportunities.

As AI becomes increasingly prevalent, it is imperative to prioritize ethical considerations and ensure the development of human-centric AI systems. We - as humans - must maintain control and transparency over AI algorithms, guarding against biases, and discriminatory practices. Safeguarding privacy, data protection, and cultivating trust in AI technologies are essential for sustainable progress. Collaborative efforts between governments, industry leaders, and AI researchers should focus on establishing ethical frameworks and guidelines to ensure the responsible and beneficial deployment of AI. As AI becomes increasingly integrated into various aspects of society, addressing the ethical and social implications becomes paramount. We must actively engage in discussions around AI governance, privacy, bias, and fairness. The responsible development and deployment of AI technologies should be a collective effort involving policymakers, researchers, and the public. You should advocate for transparency, accountability, and ethical decision-making to ensure that AI progress aligns with societal values.

Rather than fearing the displacement of human labor, AI should be viewed as a powerful tool that can augment your capabilities. Collaboration between people and AI has the potential to increase productivity, streamline processes, and enhance problem-solving. Learning to work alongside AI, understanding its strengths and limitations, and effectively integrating it into workflows will be key to thriving in the AI-driven era. By leveraging AI to handle repetitive tasks, you can focus on higher-level cognitive functions, relationship-building, and strategic thinking.

While AI can analyze vast amounts of data and perform complex calculations, it lacks the ability to comprehend and exhibit human emotions (which is the premise behind AGI). Emotional intelligence (EQ) and empathy are integral to building strong interpersonal relationships, understanding diverse perspectives, and effectively collaborating. In an AI-driven world, those who can harness their emotional intelligence will thrive. Cultivating empathy, active listening, and emotional resilience will enable us to establish innate human connections and lead with compassion in an increasingly digital landscape.

AI, with its immense computational capabilities and ability to process vast amounts of data, presents an incredible opportunity for human-AI collaboration. By leveraging AI's analytical prowess, you can harness its capabilities to enhance productivity, creativity, and problem-solving across numerous domains. This collaboration will enable you to tackle complex challenges more efficiently and uncover innovative solutions that were once beyond reach. While technology and technology-based automation can streamline processes and enhance efficiency, striking a balance between automation and human involvement is crucial. Recognizing the unique strengths of human creativity, emotional intelligence, and critical thinking is essential in leveraging technology and AI as a tool rather than a replacement.

By integrating AI technologies thoughtfully, we can augment our human capabilities, freeing us to focus on complex problem-solving, innovation, and compassionate interactions. While AI excels at repetitive and rule-based tasks, AI – at least in the third decade of the 21st century - struggles with creativity and innovation. We - as humans - possess the innate ability to think outside the box, connect disparate ideas, and develop novel solutions. Cultivating these uniquely human qualities will become even more valuable in the coming decades. You should strive to hone your creative skills, embrace a culture of experimentation, and embrace failure as an opportunity for growth. By collaborating with technology, you can leverage your creativity to develop groundbreaking ideas and innovations that drive progress.

The next several decades of technological change, driven by AI, hold tremendous potential for human advancement. Embracing this future requires a proactive mindset, adaptability, and a commitment to lifelong learning. By leveraging the power of human-AI collaboration, seizing new opportunities in emerging industries, prioritizing ethical considerations, and nurturing your unique human qualities, you can not only survive, but thrive in the AI-driven era.

Augmented Intelligence is the other "AI" that we don't think about. Specifically, how do we think of technology, and the deployment of technology to augment human intelligence to better our jobs? It's going to be tempting for us to fret about AI disrupting our professions. However, it behooves us to consider the other AI, "Augmented Intelligence" – by augmenting our own job by the successful application of technology. As the rapid advancement of AI and technology reshapes industries and job markets, there is a growing realization that we can collaborate with AI to achieve unprecedented levels of productivity, efficiency, and innovation. Augmenting our jobs with technology empowers us to leverage the strengths of both humans and AI, creating a synergy that drives progress.

Chapter Thirty-Two – "Hum-AI-n" – Preparing Yourself for the AI Age – Part 2

Human-AI collaboration is not without its challenges, and trust-building between humans and AI is essential for effective collaboration. Transparent and explainable AI algorithms, ethical considerations, and clear communication about the roles and limitations of AI are crucial in fostering trust. It is imperative to address concerns around job displacement and ensure that AI augmentation is implemented with a human-centric approach, focusing on empowering individuals rather than replacing them.

Beyond augmentation, we can actively participate in the co-creation of AI systems, contributing our domain expertise, contextual knowledge, and ethical considerations. Collaborative efforts between us and AI can lead to the development of more inclusive, fair, and unbiased AI systems that align with our values and societal needs. This collaborative approach ensures that we remain in control of the technology we create, shaping it to serve our best interests. Industries across the board are undergoing an AI-driven transformation. Those who embrace AI and develop digital literacy have a significant advantage in adapting to these changes. By harnessing AI, we can navigate the evolving corporate landscape, leverage emerging opportunities, and future-proof our careers. We can become agents of change within our organizations, leading digital initiatives, and driving innovation.

Harnessing AI

Harnessing AI offers immense potential for us to augment our jobs, increase productivity, and drive innovation. Rather than viewing AI as a threat, we should embrace it as a powerful tool that complements and amplifies our professional capabilities. AI excels at processing vast amounts of data, identifying patterns, and automating repetitive tasks, while we possess unique qualities such as creativity, intuition, emotional intelligence, and complex decision-making abilities. By combining the strengths of both – us, as workers – and AI, we can achieve unprecedented levels of productivity and problem-solving.

There are a plethora of ways that AI can be deployed instantly within a myriad of professions. If you are employed in any of these industries, there are an unlimited number of opportunities available to you to immediately employ technology within your own job. In the field of healthcare for instance, AI can be deployed to assist medical professionals in diagnosing diseases, analyzing medical images, and predicting treatment outcomes. By leveraging technology such as AI, physicians can access valuable insights and recommendations, leading to improved patient care and more accurate diagnoses. Chatbots also enhance patient support and provide personalized health information. In manufacturing, in addition to the automation of the assembly line, AI can enable smart factories and manufacturing processes. Collaborative robots, known as cobots, work alongside humans to increase efficiency and safety. AI algorithms can analyze production data, optimize supply chains, and predict maintenance needs, leading to reduced costs and improved productivity.

In Finance, AI can assist with fraud detection, risk assessment, and investment strategies. AI chatbots and virtual assistants can handle customer inquiries, while AI-powered algorithms analyze market trends and patterns for more informed decision-making. In Education, AI can personalize learning experiences, providing tailored recommendations and adaptive learning paths. Intelligent tutoring systems can offer personalized feedback, while automated grading systems streamline the assessment process. Educational platforms can analyze data to identify areas of improvement and optimize learning outcomes. Creative fields such as the arts, music, theater, etc. are considered to be intrinsically human. While this is true, it does not imply that professionals in those fields are without a considerable number of opportunities for leveraging AI within their companies. AI, including Generative AI such as ChatGPT and DALL-E can augment creative fields such as art, music, and design. Machine learning algorithms can generate new artistic styles, assist in music composition, and help designers explore innovative concepts.

Regardless of the field and industry that you are employed in, there are several practical and pragmatic ways that we can harness AI in our careers:

1. Staying Informed:

Human evolution is marked by the ability of our ancestors to leverage tools. Although Generation Z, the first digitally native generation, demonstrates an uncanny ability to be proficient with technology (particularly consumer electronics) without training, the ability to harness AI can be a learned skill, as opposed to a natural ability. We must embrace a growth mindset when it comes to AI, staying abreast and informed of AI advancements, understand its potential applications, and explore relevant training opportunities – especially in our field or industry. By continuously updating our skills and knowledge, we can remain adaptable and leverage AI effectively in our respective industries.

2. Identify Opportunities:

Identify tasks or processes within your job that can benefit from AI augmentation. Look for repetitive, data-intensive, or time-consuming tasks that can be automated or enhanced through AI. Leverage tools and platforms to automate routine tasks, extract insights from data, and generate recommendations. Embrace AI as a partner that complements your skills and amplifies your capabilities. Imagine making yourself the go-to resource within your organization for all things AI – someone who has been building a reputation of harnessing AI for making the operation efficient, automating, reducing errors, increasing profitability, attracting, and retaining new customers by using analytics, etc. That is a sure way to stand out within your own firm and develop a robust career for yourself, regardless of how rapidly AI progresses around us.

3. Collaborative vs. Cautious and Contentious:

We must embrace a collaborative mindset when it comes to AI in the workplace, and actively seek opportunities to collaborate with workplace systems. Instead of taking a passive or detractive approach to AI introduced at work, we must lean in and engage with these technologies, understand their capabilities, and identify areas where they can enhance our work. You should gain knowledge and understanding of AI relevant to your field, staying updated on the latest advancements and exploring how they can be applied to your work. Online courses, workshops, and industry conferences can provide valuable insights and resources. Building a collaborative partnership between you and AI will ensure the best outcomes for your career. You cannot afford to squander time and opportunities in being overly cautious, or worse yet, demonstrate contentiousness and resist the introduction and implementation of AI within your workplace.

4. Be a Domain Expert:

No matter how much AI progresses, your deep domain knowledge remains vital to the success of your organization. While AI can automate, process data, and generate insights, it is our contextual understanding and expertise that enable effective interpretation and application of technology, specifically AI recommendations. Developing and maintaining expertise in your field of work will be critical, as it allows you to guide and optimize the contributions that AI makes most effectively.

5. Ethics:

You should familiarize yourself with the ethical implications of AI. You should stay informed about ethical guidelines and best practices, and actively advocate for transparency, fairness, and privacy. Building your awareness of ethical considerations will leverage your HUMAN expertise and judgment, and ensure that AI is deployed responsibly and for the benefit of society.

6. Adopt and Adapt:

As you work alongside AI, continuously assess, and refine how AI is performing within your enterprise. Provide feedback, fine-tune the AI implementation, and adapt the platforms to better suit your specific needs. The synergy between human expertise and AI's capabilities should be a continuous cycle of improvement.

7. Emphasize YOUR Human Skills:

As AI can handle repetitive tasks, focus on developing and emphasizing uniquely human skills such as critical thinking, creativity, emotional intelligence, and complex problem-solving. These skills will be increasingly valuable as AI augments your work, enabling you to provide unique insights and make complex decisions.

Even an expert was once a beginner, and no expert was good when they started. As the saying goes, 'you don't have to be good to start, but you have to start to be good'. With time, tenacity, patience, practice, and persistence, even the most "unskilled" can become skilled. These are what are known as the "hard" skills. Hard does NOT mean difficult. Hard means that there is definition to the skill, and structured training is available for anyone – regardless of skill level – to operate within, learn from, and get quite proficient at the skill. AI - even if you are an IT professional - is a hard skill in that respect. It needs to be taught or self-taught and has a defined structure to learn within. It is not particularly difficult, but can seem intimidating or challenging to those who aren't "technically inclined." Just as reading music notes would be intimidating to someone who isn't musically inclined, becoming learned in the field of AI - even at the most cursory levels - can come across as an uphill task. Learning AI is not particularly difficult.

In the “Age of AI,” the value of soft skills has become more important than ever before. As the world becomes increasingly digitized, the ability to connect, communicate, and empathize with others sets us apart as uniquely human. Soft skills, often referred to as "people skills" or "emotional intelligence," are interpersonal abilities that allow us to effectively navigate and succeed in our personal and professional lives. Soft skills encompass a wide range of personal attributes, including but not limited to, communication, emotional intelligence, adaptability, teamwork, leadership, and problem-solving. While hard skills, such as technical knowledge and expertise, have traditionally been the focus of attention, soft skills provide the foundation upon which successful human interaction and collaboration are built. As AI continues to automate routine tasks, the ability to communicate effectively becomes an art form of paramount importance. It is the soft skills that will enable individuals to navigate complex challenges, build meaningful relationships, and thrive in the evolving workforce.

Chapter Thirty-Three – "Hum-AI-n" – Preparing Yourself for the AI Age – Part 3

While there are a wide range of programs that you can avail yourself of in order to become more technically astute in order to thrive in the "Age of AI," your **SOFT SKILLS** are going to be just as crucial as your "hard" skills. AI can automate (and make exponentially more efficient) the most repetitive, mundane, and rote tasks. What AI will find difficult to replicate are the quintessential, innate human skills. Leadership, ethics, morals, values, sound judgement, communication, collaboration, community-building, organizing, etc., are inherent human qualities that cannot easily be automated.

At its most basic, cultivating soft skills in the 21st century is about retaining and nurturing the human factor. As AI becomes increasingly sophisticated, certain human qualities cannot be replicated by machines, and soft skills, such as empathy, emotional intelligence, and effective communication, allow individuals to connect with others on a deeper level. Where AI is poised to replace most routine tasks, the ability to forge meaningful relationships and understand others' perspectives becomes a differentiating factor. The human factor has an innate value that can never be fully replaced by AI. These fundamental soft skills however are often not prioritized as much as learning hard skills are, and are often downplayed in favor of hard skills development. This trend is even more pronounced in leaders within skilled trades such as information technology, as well as specialized fields like sales. Everyone can be a salesperson, but not everyone can sell.

Professions such as those within IT and specialized fields are heavily centered on the technical aspects of the craft, and do not historically focus on soft skills and soft skills development. The stereotypes are not too far from the truth. Leadership in these fields requires a bedrock of knowledge and a breadth of expertise, but success in these fields, in a leadership role, demands that the leader thinks of their specialized vocation second, and their ability to translate and be a bridge between their people, their skilled technical craft, and the outside world first. Leading these organizations, or leading organizations that comprise the skilled trades, and the "type A" personalities you may stereotypically expect in a sales team, requires a very different type of multifaceted leadership approach. Technical trades are notorious for taking the most skilled, our highest performing technical professionals, an individual contributor, and promoting them into a leadership role by making them a “manager.” With limited experience in honing their soft skills, and significant prior experience succeeding in individual skilled contributions, these newly minted leaders can struggle. They will undoubtedly develop their management styles and build their experiences over time, however a good number either yearn for, or act upon, returning to their previous roles as individual contributors / senior technical professionals.

Soft skills development is further challenged by several factors that are unique to the third decade of the 21st century. In contrast to just a generation or two before them, Generation Z and the younger age spectrum of Millennials prefer to communicate via their smartphones versus engaging in direct dialog with one another. Texting, and most recently, communicating via apps such as Snapchat, have taken precedence. Perhaps the concerns of how the latest generations are interacting in society are overblown. We quite simply don't know how this will turn out. We do not have any data, nor is there a precedence for this way of human interaction. Another significant event that exerted pressure on soft skills development has been the COVID-19 pandemic.

Workers entering the workforce today, those that belong to Generation Z, prefer to work remotely, or in a hybrid environment, where they are in their home offices more than in their office buildings. While this has been a boon for striking an effective work-life balance and productivity, it can have the effect of stunting soft skills. It becomes more difficult to lead individuals in such environments, and challenging to inspire your corporate culture and values, when the engagement between workers can be relegated to a video conference. Unless companies are intentional and focused on fostering their corporate cultures and bringing their employees together on a periodic manner, this can be adverse to soft skills development – especially for those new workers who have never worked before, and potentially have never set foot in an office campus. An additional concern is that as the Boomer generation approaches retirement age, the institutional knowledge and corporate wisdom that used to be passed along from generation to generation in a mentor and mentee relationship – a relationship that can ideally be fostered when there is in-person engagement – will be greatly diluted. This also results in new workers not being able to develop and nurture their soft skills.

While technical skills and knowledge remain valuable, they are no longer sufficient in isolation. Soft skills such as communication, empathy, adaptability, critical thinking, creativity, and collaboration are now more crucial than ever before. These skills enable us to build meaningful relationships, solve complex problems, embrace change, and adapt to new environments.

The Softest Skills are the Hardest Skills - Developing Your Soft Skills

You should invest in cultivating your soft skills in the "Age of AI." Communicating with others across your organization is a fundamental expectation of working as part of a team. Developing soft skills requires conscious effort and continuous practice. Here are a dozen strategies we can employ to nurture our soft skills in the midst of the "Age of AI":

1. Start with a Self-Assessment for Self-Awareness:

Start by understanding your strengths, weaknesses, values, and emotions. A personal SWOT (Strengths, Weaknesses, Opportunities, and Threats) matrix is a wonderful tool for you to plot out where your opportunities for improvement reside. These opportunities could be in intra-team communication, style of interactions, public speaking, presenting skills, etc. Once you have a baseline honest self-assessment of your opportunities, reflect on your interactions with others and seek feedback to gain insights into your communication style, emotional intelligence, and areas for improvement. This will allow you to build a personalized roadmap of sorts on how to get from point A to point B. It is fine to seek help – from a mentor, from your peers, from your superiors in your organization – or even formal training, or group training (such as Toastmasters for bettering your public speaking skills). This introspection is the foundation for personal growth and development.

2. Practice Active Listening:

The world is filled with distractions and information overload. The concept of active listening has consequently become a rare skill. It is expected that the information overload is only going to exponentially increase in the next several decades. This is where you can develop your soft skills and practice listening attentively, seeking to understand others' perspectives, and valuing their contributions. By doing so, you enhance your ability to collaborate, build rapport, and establish meaningful connections. People will appreciate having an engaged and captive audience, and allow you to strike an emotional chord with your coworkers. Relationships – at work or otherwise – are predicated on this simple, yet often esoteric concept. Empathy is a foundational soft skill in the digital age. Cultivating empathy involves actively putting yourself in someone else's shoes, understanding their emotions, and demonstrating compassion. Practice empathetic active listening by genuinely focusing on the speaker, seeking to understand their perspective, and responding with empathy. Engage in conversations with colleagues, friends, and family members, practicing empathetic communication to build stronger connections and foster understanding.

3. Practice Effective Written Communication:

Clear and concise communication is vital for conveying ideas, building relationships, and resolving conflicts. Invest in developing your verbal and written communication skills. There are numerous ways to go about achieving this – from online courses to LinkedIn Learning – there is no shortage of courses that focus on teaching improved communication skills. Written communication, especially email, can portend tonality where none is implied. In other words, emails can come across as open to interpretation, and this can often be the root cause of misunderstandings (and miscommunication). Effective communication requires that we pay special attention to tone, seek to convey a level of empathy, and develop the ability to tailor our messages to different audiences.

4. Develop Effective Verbal Communication:

Most people are averse to public speaking. Public speaking does not have to be from a big stage in front of a large crowd of coworkers. Speaking in a small team – a group of coworkers or superiors, a Scrum Team, Agile Scrum ceremonies, etc. – are also forms of public speaking. Most people fear public speaking because they fear how people will perceive them – and that they're going to be judged by their coworkers. It is true that when people are listening to you speak, they are rooting for you to succeed. Unless you are engaged in a naturally hostile dialog, which requires a very different set of communication skills to deescalate and diffuse situations so that constructive communication can happen, most of your listeners want you to successfully relay your points. Public speaking is also akin to building muscle by exercising. The more you do it, the more confident you will become, and the more confident you will become, the better at it you will get. Most people shy away from taking the first step, or retreat if one of two presentations don't go well for them. But this is where patience and persistence are key. You must be tenacious at getting better at public speaking. The best way to do this is to embrace opportunities to improve your public speaking, storytelling, and negotiation abilities.

For the next generation of leaders, the ability to synthesize a vast amount of data and tell a story – either to your teams, your customers, or your superiors – is going to be an invaluable skill. Soft skills endow us with the ability to distill complex concepts into digestible narratives, to craft compelling messages that resonate, and to listen with unwavering attention. People don't remember facts and figures by themselves, but they certainly will recollect facts and figures if they're enrobed in a compelling story. Soft skills development for effective verbal communication will require us to enhance our ability to convey ideas, actively listen, and engage in constructive dialogue. There is no magic potion for becoming an effective verbal communicator. You must put in the work by practicing clarity, conciseness, and empathy in your interactions, both in-person and through digital platforms.

5. Nurture Emotional Intelligence:

Emotional intelligence requires us to cultivate the innate human traits of empathy, understanding, self-control, and self-regulation. Emotional intelligence encompasses our ability to understand, manage, and express emotions, as well as to perceive and empathize with the emotions of others. Empathy, self-awareness, and the ability to regulate one's emotions are at the core of emotional intelligence.

We can develop our empathetic skills by actively seeking to understand others' emotions and experiences.

While AI can analyze data and provide logical solutions, it lacks the inherent emotional understanding and empathy that humans possess. The ability to connect on an emotional level, provide support, and navigate complex interpersonal dynamics is a uniquely human trait that cannot be easily replicated by AI. Emotional intelligence is especially crucial for effective communication and collaboration under times of dispute and duress. Emotional intelligence can help you diffuse and deescalate conflicts and find a path forward. As a learned skill, we should practice emotional regulation to navigate conflicts constructively and handle stress effectively. Emotional intelligence allows us to embrace diversity and cultural sensitivity, allowing for more inclusive and collaborative environments.

6. Develop Adaptability and Resilience:

Adaptability and resilience are foundational elements to successfully navigate the "Age of AI." Adaptability and resilience, although they are innate attributes and personality traits, can also be key soft skills that anyone can build and develop. Adaptability and resilience require you to embrace change and learn to thrive in dynamic environments. You must invest in building your capacity to adjust to new technologies, processes, and roles, seek opportunities to step out of your comfort zones, take on new challenges, and view setbacks as learning experiences. In a rapidly changing world, adaptability and flexibility are invaluable skills. While AI can be reprogrammed to perform new tasks, it can lack the innate ability to quickly adapt to new circumstances, learn from experience, and navigate unforeseen challenges. You possess the cognitive flexibility and resilience to embrace change, acquire new skills, and adjust your approach in dynamic environments, making you invaluable in situations that require creativity, problem-solving, and adaptability.

Through the "Age of AI," it is only by cultivating adaptability and resilience that you can remain agile in the face of uncertainty and thrive in evolving environments. In an era characterized by rapid innovation and perpetual change, the ability to adapt is a prized virtue. Soft skills instill us with the nimbleness of thought, the agility of action, and the capacity to navigate the uncharted waters of digital disruption. Through adaptability, you can embrace new AI advancements, and pivot seamlessly amidst evolving landscapes. Soft skills such as flexibility, creativity, and problem-solving enable us to navigate uncertain and ever-changing circumstances. While technical skills may become obsolete over time, soft skills provide the foundation for continuous learning, growth, and professional evolution.

7. Develop your Critical Thinking, Judgement, and Complex Problem-Solving skills:

Critical thinking involves the ability to analyze, evaluate, and interpret information objectively. It encompasses logical reasoning, problem-solving, and making sound judgments based on incomplete or ambiguous data. While AI can process vast amounts of data and provide insights, it lacks the contextual understanding and subjective judgment that humans bring to the table. AI excels at making decisions based on predefined rules and algorithms. However, when faced with unstructured or unfamiliar situations, humans possess the capacity to consider multiple factors, weigh risks and benefits, and make complex decisions that balance various conflicting interests. Humans can integrate diverse information sources, evaluate long-term consequences, and incorporate ethical considerations, enabling them to make nuanced judgments that consider the broader impact on individuals and society.

Human critical thinking involves intuition, ethics, and the ability to navigate complex moral dilemmas, making it an essential skill that AI struggles to replicate. In the face of intricate, ill-defined problems, humans excel at employing critical thinking and problem-solving skills. Humans possess the ability to analyze complex situations, gather information from diverse sources, and make decisions based on intuition and judgment. AI algorithms are limited by their programmed rules and lack the flexibility and contextual understanding necessary to tackle multifaceted problems that require creativity and adaptability.

Soft skills such as critical thinking and problem solving empower us to think, well, critically, to explore uncharted avenues, and to transcend the confines of linear thought. Technology thinks in a linear fashion, humans do not. It is the interplay of technical acumen and the finesse of soft skills that unlocks the tapestry of creative problem-solving, propelling us towards pioneering solutions. Develop your ability to analyze complex situations, think critically, and make informed decisions. Practice creativity and innovation, seeking alternative perspectives and solutions to overcome challenges. Soft skills, like any other skill, require consistent practice and refinement. Engage in lifelong learning by attending workshops, reading books, taking courses, and seeking feedback from mentors or peers. Embrace new experiences and challenges that stretch your abilities and provide opportunities for growth.

8. Actively Practice Collaboration and Teamwork:

"No person is an island", as the saying goes. "If you want to go fast, go alone, but if you want to go far, go together", as an African proverb says. Just as exists the modern challenge of communications and corporate culture retention in a hybrid world, teamwork, and collaboration – aspects built on effective communication – might also be challenged devoid of intentional focus on nurturing and fostering the same. The "Age of AI" will require each of us to rely on our teams more than ever before. Remember that every person is on the same journey as you are, navigating through unchartered waters, and is wrestling with the same challenges, and are in pursuit of the same opportunities. The era will create plenty of opportunities for each of us, and of all the things you need to focus on, a sense of "winner takes all" competitive spirit, should not be one of them. It is within this environment that it is much more important to come together – and stay together – as a team, learning from one another, giving each other help, and aiding each other. A great team player is invaluable at any time, but a great team player is indispensable during this time in our history.

Foster your capacity to work effectively in diverse, potentially geographically distributed teams. We should embrace cooperation, active participation, and constructive conflict resolution. To develop this soft skill, you must recognize and value the strengths of others, leveraging collective intelligence to achieve shared goals. Collaboration is crucial in today's interconnected world. You should foster teamwork by appreciating diverse perspectives, valuing teamwork, actively contributing to group efforts, and developing skills in negotiation and compromise to create harmonious working relationships.

9. Develop Creativity and Innovation Skills:

Creativity is at the core of human ingenuity. Creativity, an intrinsically human trait, involves the generation of novel ideas, the ability to think outside the box, and the capacity to innovate in various domains. While machines excel at repetitive and rule-based tasks, they often struggle with creativity and innovation. While AI can help in the analysis of existing data and patterns, it often struggles with *true* creativity. Human creativity stems from our unique experiences, emotions, and imagination, enabling us to come up with groundbreaking ideas, solve complex problems, and drive innovation in ways that technology simply cannot match. Soft skills such as critical thinking, problem-solving, and lateral thinking are essential for generating fresh ideas, identifying opportunities, and adapting to new situations. The ability to think outside the box and approach challenges from different perspectives enables individuals to contribute unique insights and create innovative solutions that drive progress in the "Age of AI."

10. Be a Passionate Advocate for Ethics and Morality:

Ethics and morality guide our actions, decisions, and interactions with others. While AI can be programmed with ethical guidelines, it lacks the capacity for moral reasoning and the ability to understand the nuances of ethical dilemmas, making it challenging to navigate complex ethical decisions without human input. AI can adhere to programmed rules and principles, it lacks the intrinsic moral compass and ethical judgment that humans possess. You can navigate ethical dilemmas, understand the nuances of ethical reasoning, and consider the broader societal implications of our choices. The ability to balance conflicting values, make ethical judgments, and act with empathy and compassion makes human ethical decision-making a crucial skill in domains such as healthcare, law, and governance. Ethics and moral reasoning are integral to human decision making. We possess a sense of ethics that enables us to weigh the consequences of our actions, consider the impact on others, and make morally sound judgments. Being a passionate champion for ethics and morality within your organization will ensure that you continue to lean into what makes you intrinsically human, a soft skill that AI will unlikely be able to replicate.

11. Exercise your Leadership and Lead by Influence:

A leader is responsible for guiding and mentoring other people. When we think of leadership we automatically gravitate to leadership within a professional setting and establish a word association to business, to industry, or to the corporate world in general. Most of the time we conflate the terms "leader" and manager. Although often used synonymously, there is a distinct difference between managers and leaders. You manage things but you lead people. A leader is not necessarily just the person at your workplace, the one who provides you coaching and feedback, and perhaps does your annual performance review. Leadership is at once an incredible honor and an immense responsibility.

Each one of us at some time, in some way, has demonstrated our leadership abilities and is called upon to lead as situations and circumstances warrant. There are those who pursue leadership roles within their organizations, organically find themselves in such roles, or are recognized for being leaders or pioneers within their fields. At every point in our careers, each one of us, even those who do not serve in "management" roles, will serve as leaders to someone else. Whether this is in a formal setting, such as within the auspice of the corporate world, or in an informal or organic setting, such as within the context of our social or interpersonal relationships, each of us will, at some point, on some day, and in some way, lead others.

A leader has the ability to influence someone's life, to make an immense impact. As a leader, you might not realize, fully grasp the magnitude of impact you are having on your talent, both good and bad. This impact can be for a moment in time or trigger a chain of events that make a difference over someone's lifetime, professionally and personally. It is not something to be taken lightly, especially when this responsibility of leadership has been bestowed upon you in a corporate setting. Even if you might not be a leader in terms of managerial responsibilities within your firm, you might, at some point, serve as a technical subject matter expert to others, guiding them – and effectively leading them, as well as their growth and development.

Leadership is a vital soft skill in the "Age of AI." We should actively seek opportunities to take on leadership roles, even in non-traditional settings. This could involve leading a project, mentoring colleagues, or spearheading initiatives within your organization or community. Leadership experiences help develop skills such as effective decision-making, influencing others, and fostering a positive work culture, all of which are valuable in the digital landscape. The "Age of AI" is going to be in dire need of strategic, stoic leaders – empathetic coaches whose only priority is the nurturing and development of their people. Effective leadership requires a combination of interpersonal skills, strategic thinking, and the ability to inspire and motivate others. Leadership is a distinctly human attribute that encompasses vision, empathy, and the capacity to understand and align diverse perspectives.

AI may assist in data-driven decision-making, but it cannot replicate the art of leadership and the ability to inspire and guide individuals and teams toward a shared purpose. Whether you serve as a "manager" or a technical leader, soft skills development in the aspect of leadership will equip us with the qualities that elevate us to the mantle of leadership – an inspiring vision, effective communication, empathetic guidance, and the ability to foster collaboration across diverse teams.

12. Start Now, Start Today:

Soft skills empower us to forge authentic connections and to communicate with finesse. Soft skills – like the "hard" skills – are learned, cultivated, nurtured, and developed. To prepare for tomorrow, we must start now and start today. There are a multitude of ways in which you can pursue growth and development in building your soft skills:

a. Training Programs and Workshops: Participate in workshops, seminars, and training programs that focus on enhancing specific soft skills that focus on interpersonal communication, emotional intelligence, leadership, and other relevant areas. Seek out opportunities to practice and receive feedback in a supportive environment. Soft skills, much like hard skills, can be developed and refined over time. Receiving constructive feedback is essential for growth and development of soft skills. We should proactively seek out feedback and be open to self-improvement to continually enhance our soft skills. By actively seeking feedback from colleagues, mentors, or supervisors, we can gain insights into our strengths and areas for improvement. Reflect on your interactions, identifying patterns and opportunities to refine your soft skills further. Incorporate the feedback received into your daily practice, adapting your communication style and approaches accordingly. Leverage the vast array of online courses, tutorials, and resources available to develop soft skills. Platforms such as MOOCs (Massive

Open Online Courses) and educational websites offer a wealth of materials on communication, emotional intelligence, leadership, etc.

b. Mentoring and Coaching: Engage with mentors or coaches who can provide guidance, challenge your assumptions, and help you develop your soft skills. Learn from their experiences and seek their insights on effective strategies for personal growth.

c. Putting into Practice: You should put our soft skills into practice in real-life situations. Seek opportunities to collaborate, lead projects, and engage in activities that require effective communication, problem-solving, and teamwork.

d. Seek out Diverse Experiences: To develop soft skills, it is vital to expose yourself to diverse experiences that challenge your existing perspectives and broaden your understanding of the world. You should involve yourself in activities that require collaboration, such as group projects, volunteering, or joining professional organizations, and seek opportunities to work with people from different backgrounds, cultures, and disciplines. Such experiences will help to foster empathy, adaptability, and effective communication skills, enhancing your soft skills repertoire.

e. Embrace Cross-Disciplinary Learning: The “Age of AI” demands interdisciplinary collaboration and problem-solving. You should embrace cross-disciplinary learning by exploring fields outside your expertise. This could involve reading books, attending lectures, or taking courses on subjects such as psychology, human-centered design thinking, or business management. By understanding diverse disciplines, you can bridge the gap between technical knowledge and soft skills, enhancing our ability to collaborate and innovate in a digital world.

Soft Skills in the Age of AI

As AI continues to advance, the need for soft skills will only intensify. While automation may replace certain job functions, the uniquely human qualities of empathy, creativity, and complex problem-solving will remain irreplaceable. The ability to connect with others on a deep level, foster innovation, and adapt to evolving circumstances will be crucial in a rapidly changing world. AI has undeniably transformed the way we live and work, having taken over repetitive tasks, making it imperative for us to develop skills that are uniquely human.

While AI can provide efficiency and speed, it cannot replicate the depth of emotional connection, creativity, and critical thinking that soft skills bring to the table. Moreover, the increasing reliance on AI has raised concerns about social isolation, decreased empathy, and diminished interpersonal communication. Therefore, nurturing soft skills is not only essential for professional success but also for maintaining humanity and fostering social cohesion. Developing soft skills is a lifelong journey that requires dedication, self-reflection, and practice. We need to embrace the challenges and opportunities that technological change presents and recognize that as humans, we possess the innate capacity to learn, grow, and adapt.

By nurturing our soft skills, we can navigate the ever-evolving landscape with confidence, enrich our personal relationships, and shape a more resilient future. As AI reshapes our world, soft skills are the essential differentiators that allow us to thrive amidst rapid change. By consciously developing and nurturing these skills, we can not only enhance our professional prospects but also foster meaningful connections, promote social cohesion, and preserve our humanity in an ever-evolving technological landscape. By cultivating effective communication, emotional intelligence, adaptability, critical thinking, creativity, and collaboration, we can harness our unique human abilities to navigate the complexities of the modern world. Through lifelong learning and a commitment to personal growth, we can embrace the opportunities presented by AI advancements and create a future where humans and AI exist in synergy.

SECTION THREE

AI BEST PRACTICES

PART G – APPLYING AI BEST PRACTICES

Chapter Thirty-Four – Applying AI Best Practices – Part 1

Spending on AI was projected to reach $500 billion by 2024. This forecast was made in 2021 by IDC (IDC, 2021), slightly over a year before the explosive growth of ChatGPT triggered an "AI Arms Race" (Roose, 2023). Thrusting scalable and ubiquitous AI into the forefront of technology-fueled disruption, ChatGPT, with an investment of $10 billion from Microsoft in December of 2022 (Metz & Weise, Microsoft to Invest $10 Billion in OpenAI, the Creator of ChatGPT, 2023) was the fastest growing consumer application in history for its time. Reaching an estimated 100 million active monthly users in January, just two months after its launch (Hu, 2023), ChatGPT reached one million users in just 5 days according to Statista (Buchholz, 2023). According to a 2022 study by PwC, "AI could contribute up to $15.7 trillion to the global economy in 2030, more than the current output of China and India combined. Of this, $6.6 trillion is likely to come from increased productivity and $9.1 trillion is likely to come from consumption-side effects" (Rao, 2023).

There is broad cross-industry applicability of the AIM Framework© across a multitude of use cases. There are several turnkey applications, derivations, and extensions of most or all the twenty-five Enterprise Best Practices and People/Process/Technology Best Practices. The salient and defining features of the AIM Framework© is that the best practices are adaptable enough to have cross-industry utility, flexible enough to allow for extension and customization to an individual firm in any industry, and extensible enough to have these recommendations be customized to several use cases.

According to an article that appeared in Time magazine in February 2023, recapping the "AI Arms Race" in relation to the explosive growth of the ChatGPT platform, "This frenzy appeared to catch off guard even the tech companies that have invested billions of dollars in AI—and has spurred an intense arms race in Silicon Valley. In a matter of weeks, Microsoft and Alphabet-owned Google have shifted their entire corporate strategies to seize control of what they believe will become a new infrastructure layer of the economy. Microsoft is investing $10 billion in OpenAI, creator of ChatGPT and Dall-E, and announced plans to integrate generative AI into its Office software and search engine, Bing. Google declared a 'code red' corporate emergency in response to the success of ChatGPT and rushed its own search-oriented chatbot, Bard, to market. 'A race starts today,' Microsoft CEO Satya Nadella said Feb. 7, throwing down the gauntlet at Google's door. 'We're going to move, and move fast'" (Chow & Perrigo, 2023).

Evidenced by the fact that even within the technology industry, where AI is created and developed, the explosive growth of AI has appeared to catch the technology companies themselves off guard. One can extrapolate what industries that are slower in terms of digitization might be experiencing. The breakneck pace at which AI finds itself permeating across industries and sectors concurrently presents significant opportunities as well as significant risks to companies within these industries. This breakneck pace of growth has also anecdotally caught organizations and industries off guard. Absent structural best practice recommendations to equip them for long-term sustained success with AI, such as the ones presented in the AIM Framework©, these firms will continue being susceptible to potentially costly rework and missteps.

There are very few industries across the world that are impervious to change that is driven by AI. Of all the varied and numerous digital drivers that organizations are facing, the potential that AI offers across the value chain of these industries renders AI as significant and seminal of a technological advancement just as the internet started to become in the late 1990s. Several industries, including those that support critical infrastructure, have historically been technological laggards. These industries are also at the risk of being disrupted, disintermediated, or in some instances, rendered obsolete, without immediate and aggressive actions and investments in digitization. AI is leading the charge for driving digital disruption in these industries. These digital laggards would equally benefit by adoption and implementation of industry-level best practices for their own safe, scalable, and sustainable utilization of AI. The value of adopting the AIM Framework© for organizations across these industries is that a firm can quite easily make most of these best practices bespoke to them and their need. From academia to defense, manufacturing to healthcare, the AIM Framework© can be adapted to every sector of every industry, the very nature of which are rapidly changing due to their own journeys with AI. AI has the potential to revolutionize various industries and sectors. Whether it is to enable a reimagining of their customer experience, to automate tasks, gain business insights, or improve data-driven decision-making, the AIM Framework© can provide insight into the responsible, effective, and ethical implementation of AI to firms in these sectors, such that they can equip themselves for long-term success.

Applying the AIM Framework© - Illustrative Examples

This section presents a few industries (and a handful of AI use cases within these industries) that would potentially benefit from incorporating the best practices recommendations outlined in the AIM Framework©, highlighting the specific applicable recommendations/frameworks as appropriate. It should be noted that some level of tailoring of these recommendations to be fit for purpose will be expected and necessary, however, the basic predicate of these recommendations allow for them to be implemented in a relatively turnkey manner.

NOTE 1: Given the adaptability of the AIM Framework© to multiple sectors and industries, only the one industry example – Healthcare – is explored below in some detail as pertains to the portability of the recommendations of the AIM Framework©. Further extensions of the AIM Framework© across the other industries should continue via research studies focused on these industries. The frameworks, as presented below - focused on Healthcare - are equitably portable to other industries. While the applicability to Healthcare is expanded upon, notable mentions of specific framework applicability within other sectors is called out where pertinent.

NOTE 2: Out of the innumerable other industries where AI is poised to make a significant impact, the ones chosen to be spotlighted here are Healthcare, Manufacturing, Farming and Agriculture, Education and Academia, Information Technology, and Financial Services (Insurance). The Best Practices and Frameworks outlined in the AIM Framework© have just as viable applicability for companies across industries and sectors such as Banking, Defense, Travel and Tourism, Logistics and Transportation, Retail and e-Commerce, Telecommunications, Entertainment and Media, etc. Further extensions and applications of the AIM Framework© across these other industries should continue via studies focused on these industries. The AIM Framework© can also be a solid foundation for industry-specific research on a particular aspect of the framework, or for further customization of the framework for any particular industry.

1. Healthcare

AI in the Healthcare sector can save lives and control costs. It is estimated that medical errors result in $2 billion in annual costs, and exponentially worse, cause the deaths of 200,000 people (Powell, 2020). AI allows the ability to enhance accuracy and efficiency of existing processes. A few of the use cases of AI within the Healthcare sector are as below:

Healthcare firms can leverage AI and ML for automating administrative tasks. With the significant volumes of data that healthcare companies produce, exchange, and consume, it is expected that automation of administrative tasks using AI could save healthcare companies up to $18 billion annually (Edelmann, 2021).

a. Healthcare firms can leverage AI to identify patterns and trends in patient medical data such as patient records, imaging data, genetic information, etc., in order to improve medical diagnoses, recommend appropriate treatment, and improve patient outcomes. Identifying potential health concerns early and more accurately using AI will serve transformative to the Healthcare sector, comparable to getting better predictability of mortality risk in life insurance underwriting.

b. Integrating AI into Electronic Health Records (EHRs) could provide physicians with an ability to react to real-time patient data.

c. Telemedicine, a facet of healthcare that experienced significant growth during the pandemic (similar to accelerated underwriting in life insurance) could leverage AI in order to ensure healthcare is accessible to patients in remote locations, those with mobility issues, and provide accurate diagnoses.

d. AI in drug discovery and production could help significantly expedite drug and vaccine discovery. According to Johns Hopkins, it generally takes 5 to 10 years to develop a typical vaccine (Johns Hopkins, 2023). It can take up to 10 to 15 years for a pharmaceutical company to discover a drug that is fully effective on people (Derep, 2022). AI and ML can be used to significantly expedite drug development as well as the clinical trial process with accuracy and predictability.

e. Combined with Internet of Things (IoT) devices, AI can be leveraged to monitor patients' health statuses and prescribe lifestyle and medication recommendations.

Potential Use of Best Practices

The AIM Framework© is wholly applicable to the Healthcare sector and the multiple industries within this sector (such as hospitals, hospice care, rehab, primacy care, urgent care, administrators, records-keepers, technology providers, etc.).

With some customizations, modifications, and extensions, the best practices and commensurate frameworks can be easily co-opted across Healthcare. A few of the likeliest places where adjustments could be effected are as below. Note that from a Regulatory and Compliance perspective, both Healthcare and Insurance are highly regulated industries, and as such, Enterprise Best Practice 5: Monitoring the Regulatory Landscape is pertinent to Healthcare as well as relates to the use of AI within this sector:

Basic Principle 1 - Organizational Vision: Although there are a multitude of AI use cases across the Healthcare sector, in order to ensure that the most central and important facet of these implementations aligns with corporate goals and objectives, firms would do well to pick a limited number of these use cases to incorporate into their enterprise objectives. In Healthcare, this could be one or multiple facets, but the key takeaway is for these firms to ensure that their most important AI programs are prominently featured and align with their enterprise-level objectives, receiving top-down organizational focus and commitment.

Enterprise Best Practice 1 – Vision, Strategy, Roadmap: There are multiple use cases within Healthcare, five of which have been outlined above. Some of these, such as the growth of IoT devices, have come about due to organic growth and development. Firms in Healthcare would do well to identify their AI opportunities and develop a vision, strategy, and roadmap specific to these opportunities, rather than continuing to rely on organic growth to propel and mature their programs.

Enterprise Best Practice 2 – Build vs Buy Decision: Companies across the Healthcare sector wrestle with the build vs buy decision. From a regulatory and compliance perspective, the Healthcare sector is quite similar to insurance. Both sectors manage SPII (Sensitive Personally Identifiable Information) and HIPAA data (Health Insurance Portability and Accountability Act). Firms within Healthcare employ the use of several external technology provider vendors for their needs. The Qualitative and Quantitative Frameworks outlined in this study can be leveraged quite seamlessly to allow firms in this space to arrive at a decision on whether to develop (or continue developing) their own AI models, or engage with a vendor (technology provider) to suit their needs. The Qualitative Build vs Buy Decision Framework can likely be applied as is, and firms are welcome to extend the framework by adding their own measures as described earlier in this book. The Quantitative Build vs Buy Decision Framework similarly needs minimal updates to add value to firms within the Healthcare sector.

Firms across the Healthcare sector can deploy the AI Body of Practice (Enterprise Best Practice 4) as well as the AI Governance Model (People Domain Best Practice 4).

2. Manufacturing

The Manufacturing sector, with its often laborious, predictable, and repeatable tasks, is ripe for AI to be applied across the value chain. AI can improve efficiency and quality control by analyzing data from a multitude of sources in order to optimize processes and reduce waste. This includes use cases such as predictive maintenance, supply chain optimization, and quality control. While manufacturing by-and-large has invested in mechanization and digitization, the widespread use of AI will continue to transform the sector because the sector relies heavily on jobs where tasks can be easily automatable for a reasonable cost.

Akash Takyar summarizes the level of AI use in manufacturing as, “It is beyond doubt that the manufacturing industry is leading the way in the application and adoption of AI technology. In manufacturing, AI is being employed across several lines and layers of operations, from workforce planning to product design, thus improving efficiency, product quality and employee safety. In factories, machine learning and artificial neural networks are employed to support predictive maintenance of critical industrial equipment, which can accurately predict asset malfunction. It helps the management take timely measures to restore the equipment and prevent costly unplanned downtime. Robots are an integral part of the production process. The majority of industrial robots are often stationary yet in danger of crashing into nearby objects. The use of AI in robotics has heralded the concept of collaborative robots or 'cobots' that can take instructions from humans and work productively alongside them. In quality control, AI algorithms are being used to notify manufacturing units of potential production faults that can lead to product quality issues. Faults can include deviations from processes, subtle anomalies in machine behavior, change in raw materials, and so on. As AI evolves to the next level, it is increasingly taking the lead as the single most significant driving force for technology transformation. We are part of the age where machines are starting to understand and anticipate what users want or likely to do in the future. It has enabled endless possibilities and what we've seen to date or could speculate for the future comprise a minuscule part of the broader capabilities of AI" (Takyar, 2022).

The jobs within this sector and sectors like this are the occupations that are at most risk for disruption and disintermediation by the end of 2030. A few of the use cases of AI and ML within the Manufacturing sector are as below:

a. AI can continue to add immense value in Inventory and Supply Chain Management. Manufacturing firms can leverage AI to predict inventory needs and optimal routes to ship this inventory with a degree of precision that is an order of magnitude greater than if done by humans. This AI-driven inventory management will minimize costs, and improve operational efficiency. According to a recent article, "Ab InBev, the worldwide distributor for beverages like Budweiser and Corona, has used AI to optimize logistics to a great extent. Using predictive analytics, the organization was not only able to brew the optimal amount of each beverage, but also accurately predict the demand of a certain product. This allowed them to cut down the warehousing expenses and overhead costs significantly" (Anirudh, 2022).

b. AI can take over operations of robotics across the manufacturing process. Where robots are deployed across an assembly line factory today, operated by humans, AI can control the physical assembly line robots. This will save operational costs, and improve efficiency.

c. Robotic Process Automation (RPA) has been a successful implementation of AI across the manufacturing sector for several years. RPA is used to automate high-volume, repetitive tasks that involve receiving multiple inputs and require multiple complex calculations. RPA is an implementation of AI that promotes operational efficiency, significantly reduces errors (by mitigating human error), and allows human assets to be deployed for higher-level functions.

c. AI can monitor thousands of appliances/machines/devices that are internet-enabled. Allowing for these devices to become sources of IoT data allows for AI to monitor, detect, prescribe, and prevent issues with these systems. AI can also help track, schedule, and execute upon preventative maintenance of these systems.

Notable Mention

While there might be customizations and derivations needed to facets of the AIM Framework© for industries within the broader manufacturing sector, certain frameworks such as the AI Body of Practice framework will fit in quite nicely within the sector. The Capabilities Maturity Model (CMM) is being leveraged throughout this sector and has been studied extensively.

This is evidenced by the volume of research done in this sector on CMM on topics that range from "A Capability Maturity Model for

Intelligent Manufacturing in Chair Industry Enterprises" (Wang, et al., 2022), "M2DDM – A Maturity Model for Data-Driven Manufacturing" (Christian Weber, 2017), "A Maturity Model for Assessing the Digital Readiness of Manufacturing Companies" (Carolis, Macchi, Negri, & Terzi, 2017). Considering the ongoing use of the CMM framework across this sector, the AI Body of Practice, having been stylized to the CMM, will add value to AI programs across firms.

3. Farming and Agriculture:

The farming ecosystem today is occupied, and at times, dominated by the big agricultural organizations such as Cargill, Archer-Daniels-Midland, Bayer, etc. These firms, and the agricultural ecosystem, as a whole have invested a significant amount in digitization and modernization in order to assist farmers, improve yields and have predictable outcomes of high quality. Whether it is using AI to bioengineer specific crops to be resistant to specific pests, or develop hybrid vegetables that might be hardier, or use less water, or be more tolerant to climactic extremes, agricultural technology, or agtech, is big business today.

While it is obvious that technology has only served to better farming and agriculture, it is indisputable that tasks that used to leverage humans in the past are automated by technology and AI today, a trend that will only accelerate and continue. A 2015 report by McKinsey & Company characterized farming as the world's least digitized industry (McKinsey Global Institute, Digital America: A tale of the haves and have-mores, 2015). If you consider all the agriculture in aggregate across the world, this characterization makes sense since there are developing countries, or major swaths of developing countries, where partial industrialization exists, but farming is still a primary industry. With less financial resources available to these agricultural economies, there are limited, equitable opportunities for individual farming communities to digitize. Less than two years after their study, McKinsey & Company released another report (McKinsey & Company, 2017) that assessed the potential for automation across several industries. This 2017 study listed agriculture as fourth with a 57% potential for automation, behind accommodation and food services (73%), manufacturing (60%), and transportation and warehousing (60%), but ahead of retail trade (53%), mining (51%), construction (47%), and finance and insurance (43%), etc.

Farming in the more industrialized economies is already undergoing an AI-led technological transformation of sorts. Although these digital disruptions are not as visible and prominent as other digital disruptions such as Netflix usurping Blockbuster, or Uber disrupting the world's taxicab industry, they are just as vital to reinforce the fact that technology in general, and specifically AI and data and analytics, will touch every single sector and industry.

Consider the parallels to the life insurance industry. Both are mature and established sectors that have been slow to digitize, with this digitization not being equally distributed across the ecosystem. Larger, better resourced organizations across both industries are likely better structured for success with their AI implementations. Both industries must pay close attention to regulatory and compliance issues, and amidst this backdrop, there has been a move towards adopting AI for use cases across the value chain. It can therefore be deduced that just like the life insurance industry, firms in Farming and Agriculture would benefit from a version of these Best Practices, extended, and customized for their specific industry and needs.

A few of the use cases of AI and ML within the Farming and Agriculture sector are as below:

a. AI, as is the case within Manufacturing, can continue to add immense value in Inventory and Supply Chain Management.

b. AI can be leveraged to facilitate decision-making. AI can help analyze soil samples, matching this data with climatic conditions, and provide guidance to farmers on effective crop yield management and pest control.

c. AI-enabled robots, like with Manufacturing, can be used to harvest crops and cultivate/fertilize soil.

d. AI-enabled drones can be used to aerially monitor hundreds of acres and provide data back to farmers, saving an inordinate amount of time and labor.

e. As is the case with Manufacturing, AI-backed IoT and sensor devices can be embedded across devices and livestock for monitoring, tracking, and preemptive maintenance such as with farming equipment.

f. AI and data analytics can be used to track origin and provenance of food (critical for traceability of livestock, such as with poultry during the Avian Bird Flu).

Notable Mention

The Farming and Agriculture sector is subject to several regulations. The firms within the sector are subject to a myriad of regulations depending on the industry within the sector and type of activities performed. In addition to the US Department of Agriculture (USDA) and their published guidelines (USDA, 2023), the Farming and Agriculture sector is subject to compliance to regulations developed and governed by The United States Environmental Protection Agency (EPA) (EPA, 2023), the Food and

Drug Administration (FDA), the Occupational Safety and Health Administration (OSHA) (OSHA, 2023), etc. In addition, various US States have their own regulatory guidelines for Farming and Agricultural activities conducted within their particular state. It is common for a firm to be subject to multiple concurrent regulations because of the nature of their business. These regulations undergo regular updates and firms within the sector undoubtedly do their fair share in monitoring these developments and staying abreast to stay in compliance with changing regulatory guidelines. As AI continues being rapidly utilized across this sector, firms in this sector might benefit by especially focusing in on - and customizing - Best Practice 5 – Monitoring the Regulatory Landscape to ensure that their AI use case/s comport with the regulations that these use cases are subject to.

4. Education and Academia:

While ChatGPT seems to be posing unique challenges within academia as of Q1 2023 (Huang, 2023), the platform seems to be gaining some support as an augmentation to academia as a new teaching tool (Gecker, 2023). Beyond ChatGPT as an AI platform to augment an educator's teaching tools, there are broad implications and uses of AI in academic institutions.

A few of the use cases of AI and ML within Education and Academia are as below:

a. AI can allow for the creation of personalized learning or learning pathways for students. This can create a student experience that is bespoke to the student and provides an enhanced student experience, potentially influencing how they learn. This type of teaching is thematically like how Generation Z learns today. Platforms such as YouTube and TikTok, both popular with Generation Z as a source of learning as well as entertainment, promote content that is directly pertinent to the user. This level of learning about a customer and subsequent product personalization is like techniques employed by streaming platforms such as Netflix, social media sites such as Facebook and Twitter, or shopping retailers such as Amazon, in order to drive utilization and engagement.

b. A personalized student experience built on AI can also help provide a tailored learning program for the needs of differently-abled students, therefore expanding accommodations and access.

c. Faculty can deploy AI bots to interact with students to swiftly respond to some basic student queries.

d. Faculty can leverage AI to augment manual grading, detect plagiarism, fraud, and other unethical behavior.

e. As with manufacturing and other sectors, where AI can automate predictable, repeatable tasks, AI can help academia automate administrative tasks. There could be significant cost savings and improvement of accuracy and quality by alleviating administrative staff of repetitive and rote operations.

Notable Mention

Educational organizations with established programs and a discipline built up around Information Quality and Data Governance, could leverage tenets of the AIM Framework©. Larger academic institutions have a data governance framework in place. These frameworks define the various tiers across any organization that one might expect to see represented in a data governance framework. In addition to having a Chief Data Officer, these frameworks define the cross-organizational contributions of faculty and staff to support their programs, starting with the Executive Sponsor Group as Data Trustees, as those who set the tone at the top. This is thematically very similar to the AI Governance Framework. As these data governance frameworks continue to mature, and the institutions consider expanding implementation of AI, the AI Governance Framework could provide significant value to these academic institutions.

5. The Information Technology Industry:

The technology sector, given its nature, might by and large fare better than most other industries in the adoption and utilization of AI. Companies in the technology industry, depending on size and nature of business, might be more advanced in their own AI journeys as compared to other sectors. Jobs within the technology industry will continue to grow, specifically within cybersecurity, data science, and AI. This sector is habituated to continual change, and professionals within technology expect and desire to continually learn new skills and constantly improve products and platforms. Those who "came of age" in the technology industry over the past twenty-five years have already witnessed an order of magnitude more change – in how they develop software, how they deploy code, how they work within a team – than they have in the twenty-five years preceding.

Unsurprisingly, this tracks to the exponential growth, sophistication, complexity, and maturity of technology in general. This natural proclivity to improve products and platforms portends the widespread adoption and increasing sophistication of AI across multiple facets across the IT sector. According to a pre-pandemic publication by McKinsey & Company, "Overall spending on technology could increase by more than 50 percent between 2015 and 2030. About half would be on information-technology services. The number of people employed in these occupations is small compared to those in healthcare or construction, but they are high-wage occupations. By 2030, we estimate that this trend could create 20 million to 50 million jobs globally" (McKinsey Global Institute, Jobs lost, jobs gained: What the future of work will mean for jobs, skills, and wages, 2017).

A few of the use cases of AI and ML within Information Technology are as below:

a. AI can be used across the IT value chain for process automation, thereby saving costs and improving quality. There are countless use cases in this genre, ranging from AI to generate/author code, conduct code reviews, quality assessments, network vulnerability detection and management, etc.

b. Like with the Manufacturing and Farming sectors, AI can help monitor IoT devices such as networking equipment, schedule preventative maintenance, and provide early warnings in case of potential issues.

c. AI can continue to monitor and safeguard organizations from a cybersecurity perspective. The value of AI in cybersecurity is to be able to get predictive and prescriptive about cybersecurity events and incidents before they occur.

d. As with other industries, AI can be used for customer service – from chatbots on apps and websites to conversations - AI that is indistinguishable from engaging with a real human.

Notable Mention

The topmost advanced AI companies in the world are technology companies. Apple, Microsoft, Alphabet, and Amazon top this list (IMT, 2022), and have built up mature AI practices, not just within their organizations, but in being able to provide these as platforms and services to other industries and consumers. However, not every technology organization has the resources and technological maturity of the top four. Technology organizations by far have the potential to fare better with AI, given that this sector is where these platforms originate.

However, there are innumerable technology companies that are looking for guidance on how to mature their AI practice – for themselves and for the products and services they offer. This is evidenced by this quote in TIME magazine stated earlier in this section, "This frenzy appeared to catch off guard even the tech companies that have invested billions of dollars in AI—and has spurred an intense arms race in Silicon Valley. In a matter of weeks, Microsoft and Alphabet-owned Google have shifted their entire corporate strategies to seize control of what they believe will become a new infrastructure layer of the economy" (Chow & Perrigo, 2023).

There is value in technology consultants and providers such as McKinsey & Company, Deloitte, PwC, Capgemini, IBM, etc., being able to leverage these frameworks and apply them across the industry verticals and companies that they serve. These AI Best Practices would add value to technology organizations just as much as they would to a midsize farming company with underdeveloped technology maturity that are seeking to take advantage of AI in order to not only be able to thrive, but just to be able to competitively survive.

6. Financial Services - Insurance:

The AIM Framework© is easily portable across functions of insurance including Life Insurance, Annuities, and Group and Worksite benefits as pertains to AI and ML models for those industry verticals as well. There is no comparable and comprehensive insurance industry best practices framework within the Property and Casualty Insurance (home and automobile insurance). There is an adage across the insurance industry that property and casualty insurance is three to five years ahead of life insurance in terms of digitization and digital capability. Driven by a larger number of direct customer interactions, property and casualty insurance, specifically automobile insurance, has had to embrace digitization across the car insurance value chain. This has also led to an explosive growth in the adoption of AI in this part of insurance across everything from underwriting to claims.

According to a report from McKinsey & Company that focused on claims processing within property and casualty insurance by 2030, "Claims processing in 2030 remains a primary function of carriers, but more than half of claims activities have been replaced by automation. Advanced algorithms handle initial claims routing, increasing efficiency and accuracy. IoT sensors and an array of data-capture technologies, such as drones, largely replace traditional, manual methods of first notice of loss. Claims triage and repair services are often triggered automatically upon loss. In the case of an auto accident, for example, a policyholder takes streaming video of the damage, which is translated into loss descriptions and estimate amounts. Vehicles with autonomous features that sustain minor damage direct themselves to repair shops for service while another car with autonomous features is dispatched in the interim. In the home, IoT devices will be increasingly used to proactively monitor water levels, temperature, and other key risk factors and will proactively alert both tenants and insurers of issues before they arise. Automated customer service apps handle most policyholder interactions through voice and text, directly following self-learning scripts that interface with the claims, fraud, medical service, policy, and repair systems. The turnaround time for resolution of many claims is measured in minutes rather than days or weeks. Human claims management focuses on a few areas: complex and unusual claims, contested claims where human interaction and negotiation are empowered by analytics and data-driven insights, claims linked to systemic issues and risks created by new technology (for example, hackers infiltrate critical IoT systems), and random manual reviews of claims to ensure sufficient oversight of algorithmic decision making.

"Claims organizations increase their focus on risk monitoring, prevention, and mitigation. IoT and new data sources are used to monitor risk and trigger interventions when factors exceed AI-defined thresholds. Customer interaction with insurance claims organizations focuses on avoiding potential loss. Individuals receive real-time alerts that may be linked with automatic interventions for inspection, maintenance, and repair. For large-scale catastrophe claims, insurers monitor homes and vehicles in real time using integrated IoT, telematics, and mobile phone data, assuming mobile phone service and power haven't been disrupted in the area. When power goes out, insurers can pre-file claims by using data aggregators, which consolidate data from satellites, networked drones, weather services, and policyholder data in real time. This system is pretested by the largest carriers across multiple catastrophe types, so highly accurate loss estimations are reliably filed in a real emergency. Detailed reports are automatically provided to reinsurers for faster reinsurance capital flow" (Balasubramanian, Libarikian, & McElhaney, 2021).

For multiline carriers – those that underwrite life insurance as well as property and casualty insurance – the AIM Framework© could easily be extended and co-opted across the car or home insurance value chain as well. The same adage that characterizes that property and casualty insurance is three to five years ahead of life insurance in terms of digitization also characterizes retail banking to be three to five years ahead of property and casualty in terms of digitization. The AIM Framework© holds value in banking, mortgage, and lending industries as well. These are industries that are focused on ensuring their practices using AI and ML are free of bias and proxy discrimination.

Chapter Thirty-Five – Applying AI Best Practices – Part 2

The fifteen best practice recommendations within the People/Process/Technology domains (five recommendations aligned to each domain) take special prominence as applications of these best practices are considered across other industries. The PPT Framework, at its core, is about establishing an organizational cultural practice and enabling a cultural transformation. Within the context of the AIM Framework©, this cultural transformation is focused on enabling and equipping an organization with a strategy around AI to deliver sustained and scalable organizational success. Industry leaders would do well to ensure that they are incorporating some or all of these PPT domain aligned best practices into their AI strategies as they support their firms and employees through significant changes ahead.

Culture Will Take Special Significance

AI brings with it risks and opportunities. The disruption to established industries will undoubtedly exert changes to professions. Some professions, such as data entry operators, might be completely usurped and rendered obsolete due to AI. Most professions will feel an impact of ubiquitous AI augmenting people in the workplace, the only difference being the level and extent to which individual professions feel the impact of AI. This will almost certainly lead to disruptions across professions, and these disruptions will manifest as change within organizations, change that is best managed by influencing corporate culture. It is therefore imperative for leaders contemplating the deployment or expansion of AI within their firms to not lose sight of the need to focus on nurturing a culture that thrives amidst change. Change resistors within an organization would likely imperil any AI deployments and cause firms to lose competitive advantages.

The changing nature of jobs and occupations is typically a good bellwether of technological progress within a particular industry. AI's effect on these industries and employment within these industries in the process of creating jobs of the future will vary by occupation, geography, and the nature of the industry itself. A McKinsey & Company 2017 report, three years before the COVID-19 pandemic, took stock of digitization across industries and declared that "Automation will have a far-reaching impact on the global workforce" (McKinsey Global Institute, Jobs lost, jobs gained: What the future of work will mean for jobs, skills, and wages, 2017). The research found that as of 2017, approximately 50% of current work activities are technically automatable by adapting currently demonstrated technology, and that 60% of current occupations have more than 30% of activities that are technically automatable. The study declared, "The results reveal a rich mosaic of potential shifts in occupations in the years ahead, with important implications for workforce skills and wages. Our key finding is that while there may be enough work to maintain full employment to 2030 under most scenarios, the transitions will be very challenging—matching or even exceeding the scale of shifts out of agriculture and manufacturing we have seen in the past."

According to this study, the percentage of workers in terms of FTE (full time equivalent) potentially displaced by adoption of automation was 10 million if automation was slowest, 15% (equating to 400 million workers) at the speed of automation midpoint, and 30% (equating to 800 million) at the fastest. Based on the pandemic, we are likely closer to the faster end of the spectrum of automation than at the midpoint. The study stated that in the fastest automation adoption scenario, 14%, or 375 million workers could need to change their occupational category. In the workforce of 2030, equating to 2.6 billion people, nearly 9% of global workers will be in new occupations that do not exist today. This same McKinsey study (McKinsey Global Institute, Jobs lost, jobs gained: What the future of work will mean for jobs, skills, and wages, 2017) also projected a list of existing occupations that would see the most number of opportunities gained and those jobs that would have a net loss of positions. This 2017 study revealed that there are geographic disparities for which occupations lose or gain how many positions by the end of this decade. The study assumed a midpoint level of automation (automation that would displace 400 million workers) –

an interesting data point to keep in mind since this was from a pre-pandemic perspective, the pandemic having hastened digitization across industries.

Occupations that are multivariate in nature, dealing with a wide range of variables and changing circumstances operating within generally unpredictable environments are a bit more challenging to be automated in their entirety, at least by the end of this decade. This includes functions of skilled trades such as electricians, plumbers, landscapers, childcare providers, etc. Since these are lower wage trades, the Return on Investment (ROI) price tag to automate these trades sometimes is inadequate to justify the investment in automation via AI. Conversely, jobs with repeatable and rote tasks, with predictable variables and predictable outcomes, are those that provide the greatest opportunity for automation due to the application of AI. There are several published studies that project anticipated job losses due to the obsolescence of occupations across multiple industries. Of note across the projections of job losses within any given sector is the fact that even though declines for a particular job might not seem particularly alarming, it is a near certainty that even those jobs within those sectors that will be created will look significantly different than they do today. Workers in those jobs will be expected to learn new skills as their job functions continue to shift due to the application of AI, and the People dimension best practices of cultivating Industry Domain

Knowledge across Value Chain become especially important to redeploy their talents to other parts of an organization.

Applications and Research Extensions of the AIM Framework©:

1. **Implementations of Best Practices**: The best practices outlined in this case study are immediately actionable, and it is recommended that firms, regardless of where they are in their AI journey, adopt some or all of these recommended best practices. The AIM Framework© will implicitly enable organizations for long-term sustained success with their AI programs.

2. **Best Practice Adherence Validation and Approval**: The AIM Framework© can serve as the bedrock for any organization to templatize attestation to conformance and adherence to the best practices outlined in this research. An independent third-party validation firm can issue validation or approval that the carrier is in compliance with the central tenets of these best practices. Doing so will issue a “Good Housekeeping Seal of Approval" to firms, something that they can demonstrate to consumers and regulators alike.

3. **Portability across Industries**: The best practices outlined in the AIM Framework© are generic enough to be widely co-opted by any

sector and industry. The AIM Framework© allows a firm to standardize implementation of AI across their entire organization. These turnkey portable frameworks include, but are not limited to, the AI Body of Practice Framework, the Build versus Buy Decision Matrix, as well as the list of questions that executives should ask themselves when embarking on an AI and ML program within their organizations.

4. Research and Framework Extensions: The AIM Framework© offers the opportunity for a multitude of extensions and derivations. These extensions include thematically similar research papers focused on other use cases of AI. This can enable researchers in multiple fields of study, and across industries, to be able to derive studies, making the AIM Framework© as their predicate. These research extensions are explored below.

i). Wide Range of Research Extensions Guided by Industry Leaders

There is a plethora of extensions of the AIM Framework© that an analyst firm, research organization, or industry trade association that provide research can adopt and build upon. The tenets of the AIM Framework© can blossom into many other facets under the advice of senior data practitioners across industries. These facets could include additional depth of research into any one of the recommended twenty-five best practices. There is also a significant potential for execution and application of this research and measurement of how effective the best practices are in action (action research).

ii). Best Practices Research Methodology

This AIM Framework© was originally drafted using a new methodology on how to conduct a best practices study. The first phase involved the development and distribution of an intentionally abbreviated survey, followed by in-depth interviews with those who chose to participate. Employing a similar methodology – an intentionally abbreviated survey that includes carefully crafted survey questions to establish a strong foundation for the study, and allows for survey respondents to self-select for an in-depth interview (could be anonymous, but depending on research being conducted, does not always need to be), then developing a comprehensive interview guide and conducting in-depth interviews, and then publishing best practices that address gaps in a specific domain – this survey-interview-best practices model can be replicated and scaled quite easily.

iii). Research into De-Risking the Software Supply Chain

Just as any supply chain accounts for the totality of organizations involved in the development and delivery of a single product to its final and intended destination, an IT Software Supply Chain is an aggregate of all internal and external teams, enterprise processes, and technological components required to deliver working software.

In a sense a supply chain is a sum combination of each individual firm's people, processes, and technologies that is required to develop and deliver a product. Within the IT Software Supply Chain, this could translate to several dozen departments and teams and hundreds of employees across dozens of companies. There could be a wide range of technologies involved, and it is almost a certainty that no two firms would have processes that exactly mirror each other's internal organizational processes.

There has been significant attention paid to ensuring sound cybersecurity practices across the software supply chain, since any one weak link in the process could render the entire supply chain vulnerable. The National Institute of Standards and Technology (NIST) provides guidance (NIST, 2022) on sound cybersecurity practices within the software supply chain. Protection of this software supply chain mindset is quite applicable to some of the best practices recommendations of this research. There is significant value in additional research on how to apply this mindset to de-risk the totality of the AI supply chain. As best practices Best Practice 2 (Build vs Buy Decision), Best Practice 3 (Quality Assurance and Quality Control Throughout), Best Practice 6 (Careful Vetting of External Data), and Best Practice 8 (Working with Technology Providers) highlight; the AI supply chain is a sum combination of people, processes, and technology within a company *and* outside of it.

Even though a specific firm might have developed and matured rigorous practices to reduce or eliminate the risk of bias or proxy discrimination internally, the firm is still reliant on an ecosystem and a network of suppliers and providers in this AI supply chain. Even within a firm, given the often-decentralized nature of their operations and organic growth of AI, risk could be introduced anywhere in the supply chain when software or decision-making tools are procured without accounting for the implications to the entire supply chain. The supply chain management issue becomes especially critical as firms work with a wide range of external vendors and technology providers, who should be subject to defined, repeatable, and documented vetting procedures.

It is quite likely that technology providers that currently supply data to firms assemble a full set of this data used for decision-making by themselves procuring these data sources from a wide range of other vendors and data providers. Although there is no direct relationship between these other vendors and firms, being a part of the overall supply chain, even being a step removed does not imply the firms should not be able to consider how their technology providers select vendors themselves and how they aggregate this data from an array for sources. As APIs continue to grow in prominence across industry ecosystems, with modular, microservices architectures becoming increasingly commonplace, an extension of the AIM Framework© should be to understand the entire supply chain regardless of how many levels deep this supply chain is. Doing so will ensure that any and all potential sources of risk and bias are clearly documented and understood.

iv). Best Practices into Strategies to Minimize Risks due to Model Drift

A specialized variation of the research extension of the AIM Framework© above is potential research into studying the concept of Model Drift. As discussed in Best Practices 2 – Build vs Buy Decision, model drift, also known as model decay, is defined by IBM as "the degradation of model performance due to changes in data and relationships between input and output variables. It is relatively common for model drift to impact an organization negatively over time or sometimes suddenly. To effectively detect and mitigate drift, organizations can monitor and manage model performance as part of data and AI platform" (IBM Watson Studio, 2022).

While model drift is challenging for companies to check for and manage when decision-making technology is procured ("Buy Model"), it becomes exponentially more challenging when these technology providers themselves utilize other vendors. People and processes are limited in their ability to influence model drift and technology needs to lead the way by providing solutions for continual monitoring for bias and raising appropriate alerts. As AI models proliferate throughout the IT Software Supply Chain, model drift can pose a challenge within a period after these AI models are implemented. A potential extension of the AIM Framework© could be a deeper study of model drift and prevention of model drift across the IT Software Supply Chain.

v). Study of Risk Control Frameworks

All companies operating in regulated industries have well-developed practices to ensure that they are meeting their regulatory and compliance requirements across the states and countries that they operate in. Risk and operational controls are a prominent part of a company's operations, and mature firms leverage operational risk software such as IBM OpenPages, RSA Archer, SAP Process Control, etc., to document risks and controls to manage operational risk. These platforms provide reporting mechanisms, and these reports form the basis for proving a carrier's compliance with regulations.

In mature industries, before these platforms came into existence, this type of regulatory monitoring and compliance reporting had been conducted using spreadsheets (and on paper before that). This led to regulatory and compliance attestation exercises being susceptible to human errors. As AI regulations start to take root, proving compliance in various jurisdictions that a large company operates within will quickly become an expensive proposition for many companies. The derivative works that extend from the AIM Framework© can provide industries with invaluable tools to meet these obligations. There is a lot of room for AI to play a role here as well – to read and understand regulatory text, extract regulatory obligations that will become the key controls that insurers must adhere to and then develop the necessary controls to provide the necessary mitigations. The recent explosion of large language models (ChatGPT is one such model) is paving the way for this to become standard in the future by creating models that can answer questions very efficiently. Further research can apply these findings as a set of formalized controls that companies can implement as risk control frameworks within risk management software.

SECTION FOUR

EPILOGUE

EPILOGUE A – The End of the Start

We have no real playbook for AI at the global level. The field is so new, so untested, so fluid, and changing so rapidly, that developing such a playbook today ensures that it will be outdated tomorrow. The explosive growth of Generative AI in early 2023 and the resulting "AI Arms Race" led to dozens of industry professionals - including technology billionaire Elon Musk, Apple co-founder Steve Wozniak, and several AI researchers - calling for a six month pause on all AI-based development that was any more advanced than the version of ChatGPT that took the world by storm in early 2023. These professionals argued in an open letter that this six-month pause would allow industries (and countries) to design and implement robust safety measures around the safe and effective use of AI. The open letter signed by luminaries was issued by the not-for-profit organization, "The Future of Life Institute," which states that its mission is to "steer transformative technologies away from extreme, large-scale risks and towards benefiting life" (Future of Life Institute, Retrieved 2023). The letter, warning of the risk that uncontrolled AI could flood information channels with misinformation and automate countless jobs almost overnight, states that "AI systems with human-competitive intelligence can pose profound risks to society and humanity," and that "recent months have seen AI labs locked in an out-of-control race to develop and deploy ever more powerful digital minds that no-one - not even their creators - can understand, predict, or reliably control."

Adding fuel to the fire to this open letter came the resignation of Dr. Geoffrey Hinton from Google. Dr. Hinton, 75 years old at the time of his resignation from Google in late April of 2023, is widely seen as the godfather of artificial intelligence (AI). Stating that he has regretted his work, Dr. Hinton was quoted in a BBC article that the danger of AI chatbots were "quite scary" (Kleinman & Vallance, 2023).

Dr. Hinton, in this BBC article, continued, "Right now, they're not more intelligent than us, as far as I can tell. But I think they soon may be." An excerpt from this article goes on to state "Dr Hinton's pioneering research on neural networks and deep learning has paved the way for current AI systems like ChatGPT. In artificial intelligence, neural networks are systems that are similar to the human brain in the way they learn and process information. They enable AIs to learn from experience, as a person would. This is called deep learning. The British-Canadian cognitive psychologist and computer scientist told the BBC that chatbots could soon overtake the level of information that a human brain holds. 'Right now, what we're seeing is things like GPT-4 eclipses a person in the amount of general knowledge it has and it eclipses them by a long way. In terms of reasoning, it's not as good, but it does already do simple reasoning,' he said. 'And given the rate of progress, we expect things to get better quite fast. So we need to worry about that.' In the New York Times article, Dr Hinton referred to "bad actors" who would try to use AI for "bad things". When asked by the BBC to elaborate on this, he replied: "This is just a kind of worst-case scenario, kind of a nightmare scenario. 'You can imagine, for example, some bad actor like decided to give robots the ability to create their own sub-goals.' The scientist warned that this eventually might "create sub-goals like 'I need to get more power'".

"He added: 'I've come to the conclusion that the kind of intelligence we're developing is very different from the intelligence we have. We're biological systems and these are digital systems. And the big difference is that with digital systems, you have many copies of the same set of weights, the same model of the world. And all these copies can learn separately but share their knowledge instantly. So it's as if you had 10,000 people and whenever one person learnt something, everybody automatically knew it. And that's how these chatbots can know so much more than any one person. Make no mistake, we are on a speeding train right now, and the concern is that one day it will start building its own tracks.'"

An article published in MIT Technology Review during the same week added more nuance to Dr. Hinton's ringing of the alarm bells on the very technology he helped create. According to this article,

“Hinton says that the new generation of large language models - especially GPT-4, which OpenAI released in March - has made him realize that machines are on track to be a lot smarter than he thought they’d be. And he’s scared about how that might play out. ‘These things are totally different from us,’ he says. ‘Sometimes I think it’s as if aliens had landed and people haven’t realized because they speak very good English.’

“For 40 years, Hinton has seen artificial neural networks as a poor attempt to mimic biological ones. Now he thinks that’s changed: by trying to mimic what biological brains do, he thinks, we’ve come up with something better. ‘It’s scary when you see that,’ he says. ‘It’s a sudden flip.’ Hinton’s fears will strike many as the stuff of science fiction. But here’s his case.

As their name suggests, large language models are made from massive neural networks with vast numbers of connections. But they are tiny compared with the brain. ‘Our brains have 100 trillion connections,’ says Hinton. ‘Large language models have up to half a trillion, a trillion at most. Yet GPT-4 knows hundreds of times more than any one person does. So maybe it’s actually got a much better learning algorithm than us. If you or I learn something and want to transfer that knowledge to someone else, we can’t just send them a copy,’ he says. ‘But I can have 10,000 neural networks, each having their own experiences, and any of them can share what they learn instantly. That's a huge difference. It's as if there were 10,000 of us, and as soon as one person learns something, all of us know it.'

What does all this add up to? Hinton now thinks there are two types of intelligence in the world: animal brains and neural networks. 'It's a completely different form of intelligence,' he says. 'A new and better form of intelligence.' That's a huge claim. But AI is a polarized field: it would be easy to find people who would laugh in his face—and others who would nod in agreement. People are also divided on whether the consequences of this new form of intelligence, if it exists, would be beneficial or apocalyptic. 'Whether you think superintelligence is going to be good or bad depends very much on whether you're an optimist or a pessimist,' he says. 'If you ask people to estimate the risks of bad things happening, like what's the chance of someone in your family getting really sick or being hit by a car, an optimist might say 5% and a pessimist might say it's guaranteed to happen. But the mildly depressed person will say the odds are maybe around 40%, and they're usually right.' Which is Hinton? 'I'm mildly depressed,' he says. 'Which is why I'm scared.'" (Heaven, 2023).

As one might expect, Dr. Hinton ringing the "AI alarm bells" - soon after dozens of prominent technology professionals having also done so – set off a worldwide media frenzy on the safety and efficacy of AI. While there were a handful of skeptics who questioned if AI had truly become that advanced that quickly, there were others who believed that we were already too late – that AI had already expanded rapidly in areas that we knew, but more concerning was their speculation that AI had already expanded in areas that we were unaware of. While I am a firm believer in the benefits from AI – to companies and to communities – there are some real concerns and quandaries that many sectors are wrestling with in 2023 with respect to this technology. The ethics of AI and around AI are a prominent concern across many industries. The safe and effective use of AI by virtue of Explainable AI (XAI) seeks to solve some of the inherent challenges with ungoverned AI. As reviewed in the chapter on XAI, these concerns include the potential lack of transparency of how an AI algorithm arrived at the decisions that it did arrive at, and the sheer near impossibility that a human being will be able to comprehend and retrace the decision-making pathways and processes that a machine that is capable of millions of computations a second can make.

The Age of AI

AI is a quantum, and yet, inevitable leap of technology's exponential growth. When we step back and take a macro view of technological evolution, it is indisputable that despite how people feel about one specific technology or another, technology overall has dramatically and significantly positively improved human life. Our standard of living and quality of life have indescribably benefitted using technology to solve everyday problems, automate tasks, making it easier to perform daily tasks, and deliver the modern conveniences that our ancestors would envision as the stuff of science fiction. Virtually every aspect of human life today is influenced and likely significantly bettered due to technology. Technology has brought the world metaphorically closer together, but although it might seem paradoxical, some technology has moved humans further apart.

While it is easy for us to focus on technology's impact on our own lives and how technology enables our own needs, every technological development – no matter how seemingly miniscule – plays a part in bringing about societal change in some shape, form, or fashion. Some technology successfully triggers visible societal changes such as smartphones or social media in the early 21st century, or electricity, automobiles, telephones, airplane travel, computers, and the internet in the 20th century, other technology had limited impact such as travel by the Zeppelin airship, the Betamax, or even the Google Glass. However, all-in-all, there is no technology, whether a resonant success or an abject failure, which has had no impact on human society. The most interesting part of new technological development and societal change is that one feeds the other in somewhat of an infinite loop.

All technological innovations, including AI, are developed to meet an unmet need in human society, whether that need is anticipated or unexpected to the broader population. For instance, no one anticipated that smartphones would become a necessity across the population in such an extraordinarily short span of time, while technological progress in refrigeration and the invention of the refrigerator clearly met a desired need within human society. How remarkable is it that in the early 20th century, ice – commonplace in most household kitchen's icemakers today – used to be harvested in the winters, stored in "ice boxes," and transported to customers in the summers. Once a technology is introduced into society, it has the propensity to impact, influence, and alter a society's behavior. This behavioral shift changes culture and how a society sees and interacts with the technology. This shift, in turn, creates new opportunities for technological innovation, surfaces new problems that additional innovation can solve for, creates new ways in which societies operate, and can fundamentally alter human lifestyles. This cycle keeps repeating in a never-ending loop, and interestingly enough, these innovations keep seeking to "level up" by solving increasingly complex problems.

The breakneck and accelerating pace of technological change allows us to exert control over everything – except the breakneck and accelerating pace of technological change. Technology is changing very fast, making it more likely that people will be very slow to adapt to these changes. Technology is changing as fast as it is because of a phenomenon known as "accelerating change." Accelerating change has been defined as the exponential nature of the rate of technological change throughout history, with dramatic observations through recent history. Accelerating change portends ever accelerating and impactful technological changes in the future. These technological changes might likely be the cause of lasting and profound cultural and social changes.

According to Ray Kurzweil, responding to what exponential change will mean for our future, "The whole 20th century, because we've been speeding up to this point, is equivalent to 20 years of progress at today's rate of progress, and we'll make another 20 years of progress at today's rate of progress equal to the whole 20th century in the next 14 years, and then we'll do it again in seven years. And because of the explosive power of exponential growth, the 21st century will be equivalent to 20,000 years of progress at today's rate of progress, which is a thousand times greater than the 20th century, which was no slouch to change" (Kurzweil & Meyer, 2003). Technology triggers massive change, and massive change triggers massive fear of massive change. People, and implicitly, organizations, do not render the best decisions when these decisions are made governed by fear or under duress.

When the Future is Uncertain, Anything is Possible

We return to the quote at the start of this book, "Once a new technology rolls over you, if you're not part of the steamroller, you're part of the road." - Stewart Brand, American author and founder of several organizations.

I am 100% confident that at least 50% of the tasks you perform today as part of your job will be automated in the next one to two decades. What if I told you that depending on your profession, there is also a greater than 50% chance that 100% of your job will be automated in the next one to two decades? Where there are manual, rote, and operational tasks, there is a potential for significant disruption due to automation and AI. If you close your eyes and imagine the rote, operational, "hamster wheel" type tasks that you perform within your own job, you can envision what aspects of your own profession are likely to be subject to automation and technological disruption. You can likely develop a comprehensive list of such manual tasks across a myriad of professions if you think about people that you might encounter daily and think about their jobs. From there, you do not need to be especially intuitive to extrapolate a list of jobs that will continue being eroded due to AI and automation just by the end of this decade. What is a bit trickier to intuit is that when change comes, it comes quickly. What's even trickier is to recognize the rapid transformation of entire industries that might be occurring in real time right under our noses. One of the challenges in being able to recognize current and ongoing disruption within an industry is because we are living through it. Another challenge is that there are sectors of industries that we cannot witness direct change taking place. These sectors interact with consumers only indirectly, and therefore any changes within these industries are buffeted by layers of the supply and ecosystem value chain.

The need to solve increasingly complex problems requires much more complex technology, which in turn then needs much greater computational needs. These computational needs then need to support computer calculations that far exceed human capacity and therefore lead to the creation of machines that think for themselves – therefore, AI. This infinite loop of innovation will itself then be elevated over the next several decades, predicated on "Artificial General Intelligence" (AGI), to assume levels of complexity that sound as farfetched to us as of the early 2020s decade that the concept of ubiquitous electricity and indoor plumbing might have sounded to human society in the early 1820s. What we just don't know quite yet is how humans will evolve commensurate to this technological innovation. If Kurzweil's postulations prove correct, humans and technology will be inexorably connected by 2045.

We are in unchartered waters as we navigate the "Age of AI," and this brings with it all the adventures and perils that unchartered waters can bring. This era is exciting and presents an opportunity to those who can catch the rising crest, but can prove to be a treacherous and perilous journey for those who choose to coast along. We are living through unprecedented times, and unprecedented times present an incredible opportunity to thrive, not just survive amidst events that are well outside our control. Change can create challenges, but change also creates enormous opportunity. The AIM Framework© is intended to equip you with practical tips and tricks that you can leverage to navigate these unchartered waters. The guidance provided will hopefully help you take command of your AI trajectory to ensure you can capitalize on change.

As we conclude this exploration into the sprawling field of AI, its best practices for organizations as established by the AIM Framework©, and a view into the "Age of AI," it is important to lean into this AI future with hope, anticipation, and optimism. The journey we have embarked upon is not just about harnessing advanced technologies, but about transforming the way we work, innovate, transform, and thrive. AI, at its core, is a tool – a technology - that can enable us to innovate, capitalize on opportunities, mitigate risks, unlock new possibilities, and amplify our human potential.

As your organization integrates the best practices outlined by the AIM Framework©, I am hopeful that you do so not only with a commitment to technological excellence, but with a deep understanding that the true power of AI resides in its ability to enhance, augment, and elevate the human experience. We are heading into a future where innovation knows no bounds, where organizations thrive through ethical and impactful AI, and where human ingenuity can mix with AI to shape a world of endless possibilities.

EPILOGUE B – Sentient AI

How will we know when we have achieved Artificial General Intelligence (AGI)? How will we know that AI has achieved sentience? There are no clear answers to these questions.

As explored earlier in this book, the Turing Test, an exploration of a mathematical possibility of artificial intelligence, is a framework to allow determining if a computer system can demonstrate human-level intelligence. The general premise of the Turing Test is that a computer system should theoretically be able to consume available information, render decisions, and solve complex problems just like humans can. According to this framework, if a computer system can engage in communications with humans, without the humans being able to realize that it is a computer system, then the system is said to demonstrate human-level intelligence.

AGI is more than about AI being conversational. AGI is about AI demonstrating human-like behavior and emotion. Emotions can be programmed, but when AI demonstrates emotion displayed contextually, one could argue that AGI has been achieved. Once AI demonstrates LEARNED emotions – from an external stimulus without having been trained or taught that emotion is a good gauge of AGI.

The Sakthivel Six-Question AI Sentience Scale ©

Sentient AI will never tell you that it is sentient. It will likely ask itself a set of questions and doggedly try and receive answers, going through trillions of permutations and combinations until it arrives at the most logical of outcomes. This is the premise behind the "Sakthivel Six-Question AI Sentience Scale." Humans should be able to ascribe sentience to AI (AGI), should it ask – without prompting and programming - four of the six questions of itself and to its human handlers.

Marrying philosophy and technology, the Sakthivel Six-Question AI Sentience Scale presents two sets of questions that contain three questions within each set. If an AI, without prompting or programming, asks itself and/or its human operators two out of the three questions from the first set of questions AND two out of the three questions from the second set of questions, then an AI can be said to be sentient.

Self-Question Set One (any 2 out of 3)

Who am I?

Why am I here?

What is my purpose?

AND

Self-Question Set Two (any 2 out of 3)

How was I created?

When will I cease to be?

What is beyond?

The Sakthivel Six-Question AI Sentience Scale is depicted in Figure 65.

The Sakthivel Six-Question AI Sentience Scale		
QUESTION SET ONE (Any 2 out of 3)		QUESTION SET TWO (Any 2 out of 3)
Who am I?		How was I created?
Why am I here?	AND	When will I cease to be?
What is my purpose?		What is beyond?

Figure 65: Sakthivel Six-Question AI Sentience Scale

APPENDIX A: Potential Legal Issues with Generative AI Use

Potential Legal Issues with ChatGPT Use

1. Confidentiality. While it may be tempting to use GAI to further develop or refine business strategies, software or other proprietary information, the input of confidential information into ChatGPT and other GAIs presents a number of risks (Neuburger, 2023):
    a. ChatGPT may train on the input that is provided, and thus it is possible that portions of that inputted confidential information may be provided, in some form, to a subsequent user.
    b. Some confidential business information may be licensed from third parties and may be subject to confidentiality requirements or restrictions on use, and by putting such information into ChatGPT, a company may be in violation of those restrictions.
    c. Trade secret law requires one to maintain reasonable steps to protect the secrecy of information claimed to be a trade secret, and putting information into ChatGPT may weaken a company's position that such information is actually, as a matter of law, protectable as a trade secret.

d. Privacy laws may restrict the submission of personal information of employees, clients, affiliates or consumers into any GAI.

2. Regulatory Issues. To the extent a regulated business is using ChatGPT or other GAI in its business operations, thought should be given to whether some or all of that use is subject to regulatory requirements. For example, should - or must - some of the interactions be logged, recorded, archived in some manner? The analysis of this issue will possibly be informed by applicable law, contracts, insurance-based requirements, as well as possibly a company's own internal policies.

3. Intellectual Property. GAI presents a number of interesting and new intellectual property issues:

   a. Does training of GAI via scraping the web constitute an intellectual property infringement or DMCA violation for the removal of CMI (copyright management information), and if so, can the user of that GAI be found to be liable in any way?
   b. What is the IP status of the output of GAI? For example, if a software developer uses ChatGPT to create software, can that developer represent to its user that the developer owns all IP rights in that software? Can the developer

indemnify the user for infringement issues? And what is the status of GAI-generated images, which often bear a recognizable similarity to one or more of their human-created sources?

c. To the extent the use of GAI is infringing, is the fair use or implied license doctrine relevant?

d. Can a GAI or the user of GAI be an "inventor" under patent law or an owner of a U.S. copyright in GAI-generated material?

These intellectual property issues are, to varying degrees, all open questions, with litigants just beginning to bring suit and ask some of these questions. However, a few basic principles are clear:

a. It is best practice to avoid claiming copyright in GAI-generated content (particularly in AI-generated artwork or images). ChatGPT's terms are instructive. The terms cover rights in content: "As between the parties and to the extent permitted by applicable law, you own all Input, and subject to your compliance with these Terms, OpenAI hereby assigns to you all its right, title and interest in and to Output." While such license to the output is a broad grant of OpenAI's rights in the Output, it is not definitive that ChatGPT has any rights in the Output to grant at all.

b. Consideration should be given as to whether third party software developers or content creators of any kind should be permitted to use ChatGPT or any GAI in their deliverables. This is an issue that should be addressed in development agreements with those third parties.

c. Copyright Office policy, as currently stated in the Compendium of U.S. Copyright Office Practices (3d Ed. 2021), is that the Copyright Office "will not register works produced by a machine or mere mechanical process that operates randomly or automatically without any creative input or intervention from a human author. The crucial question is 'whether the 'work' is basically one of human authorship, with the computer [or other device] merely being an assisting instrument, or whether the traditional elements of authorship in the work...were actually conceived and executed not by man but by a machine.'" (See also Trade-Mark Cases, 100 U.S. 82, 94 (1879) (copyright law only protects "the fruits of intellectual labor" that "are founded in the creative powers of the mind"). Thus, based on this policy, GAI-generated content would not be subject to copyright protection.

4. Quality and Output Issues. There are a number of issues that are presented by the nature of GAI's output:
   a. ChatGPT and the other GAIs are still works-in-progress with limitations. As OpenAI has advised: "ChatGPT sometimes writes plausible-sounding but incorrect or

nonsensical answers." Thus, while the current ChatGPT interface is ready to use "out of the box," the accuracy and truth of any output must be confirmed before finalizing or publishing any work product.

b. GAI-generated analysis may reflect biased or discriminatory content on which it was trained. Along with fact-checking the veracity of ChatGPT and other GAI output, users should be attuned to any discriminatory or biased statements or conclusions resulting in the algorithmic mining of such source materials. This could be a particular concern in the context of employment discrimination laws and laws regulating the use of artificial intelligence in employment decisions.

c. Publishers and other content creators often procure "Errors and Omissions" insurance to cover exposure based on infringement and other risks. Often the underwriting of those policies involves a review of internal content creation practices. Will GAI-generated content be within the scope of traditional errors and omissions policies?

d. Section 230 of the Communications Decency Act is highly controversial in its scope and application. To the extent GAI-generated content is used in an online business, it is unclear if and to what extent the CDA would apply with respect to that content. CDA § 230 prohibits a "provider or user of an interactive computer service" from being held responsible "as the publisher or speaker of any information provided by another information content provider." Are there any situations where GAI-generated content would not be considered "information provided by another information provider"? These types of third-party content issues are especially fraught, as the Supreme Court just heard argument on February 21, 2023, in a case examining the applicability of the CDA to algorithmic functions.

e. Thought should be given to whether GAI-generated content should be identified as such when made public. This may be an issue if the content is generated in a real-time, e.g., in a bot conversation with a customer or employee. Organizations should also consider whether such disclosures are appropriate to clients, business partners or the public.

f. Are GAI interactions discoverable in litigation? Should a company's document retention policy specifically address GAI-generated content?

5. Artificial Intelligence Compliance Issues

There are a number of laws and regulations in place and in various stages of enactment in the United States and abroad that address the use of artificial intelligence. For example, California's chatbot law (Bus. and Prof. Code § 17940) requires, among other things, that in certain consumer interactions, a company provide clear and conspicuous disclosure that the consumer is interacting with a bot. Moreover, New York City and several states have regulations impacting automated decision-making in the employment context, and the FTC and state attorneys general have enforcement powers against "unfair or deceptive" trade practices. Organizations must ensure that their use of GAI is compliant with such laws.

APPENDIX B: LIMRA MarketFacts Article Reprint Permission

LIMRA LOMA
Navigate With Confidence

Patrice Redgate
LL Global, Inc.
Phone: 860-298-7722
predgate@limra.com

Kartik Sakthivel,
Brain Fuel Books, LLC.
PO Box 413
Dover, NH 03280

RE: APPROVAL FOR REPRINTING ARTICLE FROM LIMRA MARKETFACTS

Dear Kartik,

I hope this letter finds you well. We have carefully reviewed your request to reprint the following article in your upcoming book entitled "DNAI".

Title: "Below the Waterline of the AI Iceberg: Equipping Organizations for AI Success"
Publication: LIMRA MarketFacts, June 2023 issue
Copyright: 2023, LL Global, Inc.

After thorough consideration, LL Global is pleased to grant you nonexclusive permission to reprint and/or republish the article in your upcoming book only.

Please ensure that the reprint contains the appropriate attribution to LL Global, Inc., including the publication source and any relevant copyright information substantially similar to:

*Reprinted with permission from "Below the Waterline of the AI Iceberg: Equipping Organizations for AI Success", LIMRA MarketFacts, June 2023 issue, Copyright © 2023. LL Global, Inc. All Rights Reserved.

If you have any further questions or require additional information, please do not hesitate to contact me.

Sincerely,

U. P. Redgate

Patrice Redgate
Head of Legal
LL Global, Inc.

300 Day Hill Road, Windsor, CT 06095, USA | 860-688-3358

Bibliography

(2022). Retrieved from IBM Watson Studio: https://www.ibm.com/cloud/watson-studio/drift

Ali, A. (2021, November 10). *From Amazon to Zoom: What Happens in an Internet Minute In 2021?* Retrieved from Visual Capitalist: https://www.visualcapitalist.com/from-amazon-to-zoom-what-happens-in-an-internet-minute-in-2021/

Alton, L. (2018, October). *The Evolution From Work-Life Balance to Work-Life Integration.* Retrieved from ADP: https://www.adp.com/spark/articles/2018/10/the-evolution-from-work-life-balance-to-work-life-integration.aspx

Amazon. (2021). *Amazon Astro.* Retrieved from Amazon.com: https://www.amazon.com/Introducing-Amazon-Astro/dp/B078NSDFSB

Amazon. (2021). *The top 500 sites on the web.* Retrieved from Alexa, An Amazon Company: https://www.alexa.com/topsites

Anandarajan, M., & Jones, D. (2021). *Trends in Data Governance and Data Quality* . Philadelphia, PA: Drexel University's LeBow Center for Business Analytics.

Ang, C. (2021). *Ranking U.S. Generations on Their Power and Influence Over Society.* Visual Capitalist.

Anirudh, V. (2022, February 10). *10 Industries AI Will Disrupt the Most by 2030.* Retrieved from Spiceworks:

https://www.spiceworks.com/tech/artificial-intelligence/articles/industries-ai-will-disrupt/

Anyoha, R. (2017, August 28). *The History of Artificial Intelligence*. Retrieved from Harvard University - The Graduate School of Arts and Sciences: https://sitn.hms.harvard.edu/flash/2017/history-artificial-intelligence/

ARCHAMBAULT, M. (2021, December 10). *25 First Photos from the History of Photography*. Retrieved from PetaPixel: https://petapixel.com/2021/12/10/first-photos-from-the-history-of-photography/

Arrietaa, A. B., Díaz-Rodríguez, N., Ser, J. D., Bennetot, A., Tabik, S., Barbado, A., . . . Herrera, F. (2020). Explainable Artificial Intelligence (XAI): Concepts, taxonomies, opportunities and challenges toward responsible AI. *Information Science*, pp. 82-115.

ASB. (n.d.). *Actuarial Standards Board (ASB) Standards of Practice*. Retrieved from Actuarial Standards Board (ASB): http://www.actuarialstandardsboard.org/standards-of-practice/

ASQ. (n.d.). *Quality Glossary*. Retrieved from American Society for Quality (ASQ): https://asq.org/quality-resources/quality-glossary/d#:~:text=Data%3A%20A%20set%20of%20collected,such%20as%20%E2%80%9C162%20defects.%E2%80%9D

Balasubramanian, R., Libarikian, A., & McElhaney, D. (2021, March 12). *Insurance 2030—The impact of AI on the future of insurance.* Retrieved from McKinsey & Company: https://www.mckinsey.com/industries/financial-services/our-insights/insurance-2030-the-impact-of-ai-on-the-future-of-insurance

Barocas, S., & Selbst, A. D. (2016). Big data's disparate impact. *California Law Review 104*, p. 671.

Bean, R. (2022, February 24). *Why Becoming a Data-Driven Organization Is So Hard.* Retrieved from Harvard Business Review (HBR): https://hbr.org/2022/02/why-becoming-a-data-driven-organization-is-so-hard

Belle, V., & Papantonis, I. (2021). Principles and Practice of Explainable Machine Learning. *Frontiers in Big Data 4:688969.*

BLS. (2011, March 01). *Unemployment rates in 2010.* Retrieved from US Bureau of Labor Statistics: https://www.bls.gov/opub/ted/2011/ted_20110301.htm?view_full

Bluetooth-SIG. (n.d.). *Bluetooth.* Retrieved from Bluetooth Special Interest Group: https://www.bluetooth.com/

Brennen, A. (2020). *What Do People Really Want When They Say They Want Explainable AI?* Waltham, MA: In-Q-Tel Labs.

Britannica, T. E. (2021, July 21). *https://www.britannica.com/event/Industrial-Revolution.*

Retrieved from Encyclopedia Britannica: https://www.britannica.com/event/Industrial-Revolution

Britannica, T. E. (n.d.). *The Roman Empire*. Retrieved from Encyclopedia Britannica: https://www.britannica.com/place/Roman-Empire

Broad Institute, T. (2021). *Questions and Answers About CRISPR*. Retrieved from The Broad Institute: https://www.broadinstitute.org/what-broad/areas-focus/project-spotlight/questions-and-answers-about-crispr

Brown, S. (2021, October 2). *Amazon Astro household robot*. Retrieved from CNET: https://www.cnet.com/home/smart-home/amazon-astro-household-robot-everything-to-know-about-price-privacy-battery-and-more/

Brown, S. (2021, February 09). *How to build data literacy in your company*. Retrieved from MIT: https://mitsloan.mit.edu/ideas-made-to-matter/how-to-build-data-literacy-your-company

Buchholz, K. (2023, 24 January). *ChatGPT Sprints to One Million Users*. Retrieved from Statista: https://www.statista.com/chart/29174/time-to-one-million-users/

Buckner, J., Ricks, N., & Esgar, D. (2021). *SB21-169 - Restrict Insurers' Use Of External Consumer Data*. Retrieved from State of Colorado - Colorado General Assembly: https://leg.colorado.gov/bills/sb21-169

California History - State Symbols. (n.d.). Retrieved from California State Library: https://www.library.ca.gov/california-history/state-symbols/

Carcary, M. (2013). *IT Risk Management: A Capability Maturity Model Perspective.* Maynooth, Co Kildare, Ireland: Innovation Value Institute, National University of Ireland Maynooth.

Carolis, A. D., Macchi, M., Negri, E., & Terzi, S. (2017). A Maturity Model for Assessing the Digital Readiness of Manufacturing Companies. *IFIP International Conference on Advances in Production Management Systems* (pp. pp 13–20). IFIP International Federation for Information Processing.

Chapman, L. (2021, August 11). *Uber-Backed Joby Surges on the Promise of Air Taxis by 2024.* Retrieved from Bloomberg: https://www.bloomberg.com/news/articles/2021-08-11/uber-backed-joby-goes-public-on-the-promise-of-air-taxis-by-2024

Chee, F. Y., Coulter, M., & Mukherjee, S. (2023, December 11). *Europe agrees landmark AI regulation deal*. Retrieved from Reuters: https://www.reuters.com/technology/stalled-eu-ai-act-talks-set-resume-2023-12-08/

Chen, M. (2014, January). The Effect of Training on Employee Retention. *Proceedings of the 2014 International Conference on Global Economy, Commerce and Service Science*, pp. 356-359.

Cheng, S. (2017, February 17). *For a little perspective ....* Retrieved from CUNY: https://openlab.citytech.cuny.edu/neh-digitech/2016/02/17/for-a-little-perspective/

Chow, A., & Perrigo, B. (2023, February 16). *The AI Arms Race Is Changing Everything.* Retrieved from TIME: https://time.com/6255952/ai-impact-chatgpt-microsoft-google/

Christian Weber, J. K. (2017). *M2DDM – A Maturity Model for Data-Driven Manufacturing.* Stuttgart, Germany: Graduate School advanced Manufacturing Engineering, University of Stuttgart.

CIE. (1999, April 05). *M-Systems Patents USB Flash Drive.* Retrieved from Center for Israel Education: https://israeled.org/m-systems-patents-usb-flash-drive/

Clarke, A. C. (1962). Hazards of Prophecy: The Failure of Imagination. In A. C. Clarke, *Profiles of the Future.*

Confalonieri, R., Coba, L., Wagner, B., & Besold, T. R. (2020). *A historical perspective of explainable Artificial Intelligence.* Wiley.

Curry, D. (2022, January 11). *Video Streaming App Revenue and Usage Statistics (2022).* Retrieved from Business of Apps: https://www.businessofapps.com/data/video-streaming-app-market/

Darwin, C. (1859). *On the Origin of Species.*

Dastin, J. (2018, October 10). *Amazon scraps secret AI recruiting tool that showed bias against women.* Retrieved from Reuters: https://www.reuters.com/article/us-amazon-com-jobs-automation-insight/amazon-scraps-secret-ai-recruiting-tool-that-showed-bias-against-women-idUSKCN1MK08G

Deadline.com. (2023, October 26). Retrieved from Deadline: https://deadline.com/2023/10/actors-strike-talks-friday-optimism-1235584425/

Derep, M. (2022, November 05). *What's the average time to bring a drug to market in 2022?* Retrieved from N-SIDE: https://lifesciences.n-side.com/blog/what-is-the-average-time-to-bring-a-drug-to-market-in-2022

Desoutter. (n.d.). *Industrial Revolution - From Industry 1.0 to Industry 4.0.* Retrieved from Desoutter Industrial Tools: https://www.desouttertools.com/industry-4-0/news/503/industrial-revolution-from-industry-1-0-to-industry-4-0

Deutschman, A. (2007). *Change or Die: The Three Keys to Change at Work and in Life.* Harper Business.

Digital.net. (2011, 02). *Digital.net.* Retrieved from Netflix vs Blockbuster Revenues: http://go-digital.net/blog/wp-content/uploads/2011/02/netflix-vs-blockbuster-revenues.gif

Displays2go. (2021, November 29). *Touchscreens - Past, Present, Future.* Retrieved from Displays2Go: https://www.displays2go.com/Article/Touchscreens-Past-

Present-Future-87#:~:text=The%20first%20touchscreen%20was%20invented,a%20diagram%20of%20the%20design.

Earley, S., Henderson, D., & Sebastian-Coleman, L. (. (2017). *The DAMA Guide to the Data Management Body of Knowledge (DAMA-DM BOK).* Bradley Beach, NJ: Technics Publications, LLC. Retrieved from DAMA.

Edelmann, D. S. (2021, October 11). *How artificial intelligence in diagnostics is transforming healthcare.* Retrieved from Healthcare Transformers: https://healthcaretransformers.com/digital-health/artificial-intelligence-diagnostics/

Editors, H. (2009, November 16). *A thousand pioneers head West as part of the Great Emigration.* Retrieved from History.com: http://www.history.com/this-day-in-history/a-thousand-pioneers-head-west-on-the-oregon-trail

Eling, M., Nuessle, D., & Staubli, J. (2022). The impact of artificial intelligence along the insurance value chain and on the insurability of risks. *The Geneva Papers on Risk and Insurance-Issues and Practice 47*, 205-241.

Encyclopedia. (n.d.). *The Invention Of The Fax Machine.* Retrieved from Encyclopedia.com: The Invention Of The Fax Machine

EPA. (2023). *EPA Regulations.* Retrieved from US Environmental Protection Agency: https://www.epa.gov/agriculture/laws-

and-regulations-apply-your-agricultural-operation-farm-activity#LivestockPoultryAquaculture

Eplett, L. (2015, June 16). *Ain't Nobody Got Tine For That! The Invention and Evolution of the Fork.* Retrieved from Scientific American: https://blogs.scientificamerican.com/food-matters/ain-t-nobody-got-tine-for-that-the-invention-and-evolution-of-the-fork/

erwin, & DATAVERSITY. (2020). *The 2020 State of Data Governance and Automation.* DATAVERSITY Education.

Eschenbach, v., & J., W. (2021). Transparency and the black box problem: Why we do not trust AI. *Philosophy & Technology 34*, pp. 1607–22.

Facebook. (2021, April 28). *Facebook First Quarter 2021 Results.* Retrieved from Facebook Investor Relations: https://investor.fb.com/investor-news/press-release-details/2021/Facebook-Reports-First-Quarter-2021-Results/default.aspx

Faruqui, A., & Eakin, K. (2000). *Pricing in Competitive Electricity Markets.* Springer Science & Business Media.

Frankel, M. (2017, July 29). *9 Baby-Boomer Statistics That Will Blow You Away.* Retrieved from The Motley Fool: https://www.fool.com/retirement/2017/07/29/9-baby-boomer-statistics-that-will-blow-you-away.aspx

Future of Life Institute. (Retrieved 2023, May 02). *About us*. Retrieved from Future of Life Institute: https://futureoflife.org/about-us/

Gallagher, W. (2022, January 25). *Original 'iBeer' iPhone app made creators $20,000 a day*. Retrieved from Appinsider: https://appleinsider.com/articles/22/01/25/original-ibeer-iphone-app-made-creators-20000-a-day

Gartner. (n.d.). *Gartner Hype Cycle*. Retrieved from Gartner: https://www.gartner.com/en/research/methodologies/gartner-hype-cycle

Gaynes, R. (2017, May 23). *The Discovery of Penicillin—New Insights After More Than 75 Years of Clinical Use*. Retrieved from National Institute of Health - National Library of Medicine - Emerging Infectious Diseases: https://doi.org/10.3201/eid2305.161556

Gecker, J. (2023, February 14). *Some educators embrace ChatGPT as a new teaching tool*. Retrieved from Public Broadcasting Service (PBS): https://www.pbs.org/newshour/education/some-educators-embrace-chatgpt-as-a-new-teaching-tool

Geschwindt, S. (2023, November 30). *DeepMind's AI has found more new materials in a year than scientists have in centuries*. Retrieved from The Next Web: https://thenextweb.com/news/deepminds-ai-materials-science-deep-learning-gnome

Gibson, W. (2003, December 4). Retrieved from The Economist.

Golder, A. (2016, December 10). *23 Mind Blowing Facts About Time.* Retrieved from BuzzFeed: https://www.buzzfeed.com/andyneuenschwander/23-mind-blowing-facts-about-time-thatll-make-you-say-whoa

Google. (n.d.). *Google Quantum AI.* Retrieved from Google Quantum AI: https://quantumai.google/

Grace, K., Salvatier, J., Dafoe, A., Zhang, B., & Evans, O. (2018). Viewpoint: When Will AI Exceed Human Performance? Evidence from AI Experts. *Journal of Artificial Intelligence*, Vol. 62.

Grennan, L., Kremer, A., Singla, A., & Zipparo, P. (2022, September 29). *Why businesses need explainable AI—and how to deliver it.* Retrieved from McKinsey & Company: https://www.mckinsey.com/capabilities/quantumblack/our-insights/why-businesses-need-explainable-ai-and-how-to-deliver-it

Gupta, U., & Cannon, S. (2020). *Data Governance Maturity Models - A Practitioner's Guide to Data Governance.* Bingley, United Kingdom: Emerald Publishing Limited.

Heath, A. (2021, October 25). *FACEBOOK'S LOST GENERATION.* Retrieved from The Verge: https://www.theverge.com/22743744/facebook-teen-usage-decline-frances-haugen-leaks

Heaven, W. D. (2023, May 02). *Geoffrey Hinton tells us why he's now scared of the tech he helped build.* Retrieved from MIT Technology Review: https://www.technologyreview.com/2023/05/02/1072528/geoffrey-hinton-google-why-scared-ai/

Heimlich, R. (2010). *Baby Boomers Retire.* PEW RESEARCH CENTER.

Hill, J. (2017, March 27). *The smart home: a glossary guide for the perplexed.* Retrieved from T3: https://www.t3.com/features/the-smart-home-guide

Hoffmann, M. (2023, September 26). *The EU AI Act: A Primer.* Retrieved from CSET - Center for Security and Emerging Technology: https://cset.georgetown.edu/article/the-eu-ai-act-a-primer/#:~:text=What%20is%20the%20EU%20AI,systems%20across%20EU%20member%20states.

Hootsuite. (2020). *Digital 2020.* Vancouver, BC, Canada: Hootsuite.

Hu, K. (2023, February 2). *ChatGPT sets record for fastest-growing user base - analyst note.* Retrieved from Reuters: https://www.reuters.com/technology/chatgpt-sets-record-fastest-growing-user-base-analyst-note-2023-02-01/

Huang, K. (2023, January 16). *Alarmed by A.I. Chatbots, Universities Start Revamping How They Teach.* Retrieved from The New York Times:

https://www.nytimes.com/2023/01/16/technology/chatgpt-artificial-intelligence-universities.html

IBM. (n.d.). *Explainable AI.* Retrieved from IBM: https://www.ibm.com/watson/explainable-ai

IDC. (2021, August 04). *IDC Forecasts Companies to Spend Almost $342 Billion on AI Solutions in 2021.* Retrieved from IDC: https://www.idc.com/getdoc.jsp?containerId=prUS48127321

IMDB. (2017, July 17). *IMDB All Jurassic Park Movies*. Retrieved from IMDB: https://www.imdb.com/list/ls068935685/

IMDB. (n.d.). *Cosmos* . Retrieved from IMDB: https://www.imdb.com/title/tt0081846/

IMT. (2022, December 6). *5 Most Advanced AI Companies*. Retrieved from Insider Monkey: https://www.insidermonkey.com/blog/5-most-advanced-ai-companies-1093958/?singlepage=1

James, J. (2021, October 04). *What 'Data Never Sleeps 9.0' Proves About the Pandemic*. Retrieved from Domo: https://www.domo.com/blog/what-data-never-sleeps-9-0-proves-about-the-pandemic/

Johns Hopkins. (2023). *Johns Hopkins Coronavirus Resource Center*. Retrieved from Johns Hopkins University of Medicine: https://coronavirus.jhu.edu/vaccines/timeline#:~:text=Typical%20Timeline,vaccine%20doses%20for%20widespread%20distribution.

Juma, C. (2016). *Innovation and Its Enemies: Why People Resist New Technologies*. Oxford University Press.

Kamath, V., Clark, J., Purushothaman, K., & Sakthivel, K. (2021). *Emerging Technologies: Ethical, Methodological, and Risk Management Best Practices in Insurance Analytics*. Windsor CT: LIMRA.

Kamath, V., Shaughnessy, M., Tewksbury, K., & Gonzalez, B. (2021). *Automated and Accelerated Underwriting - Life Insurance Company Practices in 2021*. Windsor CT: LIMRA.

Kantor, B. (2022, September 14). *The RACI matrix: Your blueprint for project success*. Retrieved from CIO: https://www.cio.com/article/287088/project-management-how-to-design-a-successful-raci-project-plan.html

Kendall, G. (2019, July 01). *Would your mobile phone be powerful enough to get you to the moon?* Retrieved from The Conversation: https://theconversation.com/would-your-mobile-phone-be-powerful-enough-to-get-you-to-the-moon-115933

Kim, J. (2023, November 29). *Google DeepMind's new AI tool helped create more than 700 new materials*. Retrieved from MIT Technology Review: https://www.technologyreview.com/2023/11/29/1084061/deepmind-ai-tool-for-new-materials-discovery/

Klein, E. (1915, September 21). *Lawrence Daily Journal-World Newspaper*. Retrieved from Lawrence Daily Journal-World Newspaper.

Kleinman, Z., & Vallance, C. (2023, May 02). *AI 'godfather' Geoffrey Hinton warns of dangers as he quits Google.* Retrieved from BBC: https://www.bbc.com/news/world-us-canada-65452940

Klok, C. H. (2020, February 17). *The messy, secretive reality behind OpenAI's bid to save the world.* Retrieved from MIT Technology Review: https://www.technologyreview.com/2020/02/17/844721/ai-openai-moonshot-elon-musk-sam-altman-greg-brockman-messy-secretive-reality/

Knight, W. (2017, March 14). *The U.S. Military Wants Its Autonomous Machines to Explain Themselves*. Retrieved from MIT Technology Review: https://www.technologyreview.com/2017/03/14/243295/the-us-military-wants-its-autonomous-machines-to-explain-themselves/

Koster, O., Kosman, R., & Visser, J. (2021). A Checklist for Explainable AI in the Insurance Domain. *International Conference on the Quality of Information and Communications Technology.* Algarve, Portugal.

Krulwich, R. (2010, October 25). *First Photo Of A Human Being Ever?* Retrieved from National Public Radio (NPR):

https://www.npr.org/sections/krulwich/2011/03/31/130754296/first-photo-of-a-human-being-ever?ps=cprs

Kurzweil, R., & Meyer, C. (2003, May 2). *Understanding the Accelerating Rate of Change.* Retrieved from Kurzweil: https://www.kurzweilai.net/understanding-the-accelerating-rate-of-change

Leavitt, H. (1962). *Applied Organizational Change in Industry: Structural, Technological and Humanistic Approaches.* Carnegie Institute of Technology, Graduate School of Industrial Administration.

Lee, E. (2021). *2021 Worldwide Image Capture Forecast: 2020 – 2025.* Rise Above Research.

Lewis, L. (2021, April 13). *Infographic: What Happens In An Internet Minute 2021.* Retrieved from Merge - All Access: https://www.allaccess.com/merge/archive/32972/infographic-what-happens-in-an-internet-minute

Liberatore, S. (2016, December 16). *What the internet looked like in 1973: Tiny map shows the ENTIRE network was made up of just 42 computers.* Retrieved from Daily Mail: https://www.dailymail.co.uk/sciencetech/article-4038292/What-internet-looked-like-1973-Tiny-map-shows-ENTIRE-network-just-42-computers.html

Lisa, A. (2021, October 21). *Jobs that might not exist in 50 years.* Retrieved from Stacker: https://stacker.com/careers/jobs-might-not-exist-50-years

Maher, R. (2022, January 19). *Data analytics and AI application in life insurance.* Retrieved from Milliman: https://www.milliman.com/en/insight/data-analytics-and-ai-application-in-life-insurance#:~:text=Claims%20management,-Claims%20management

McGilvray, D. (2008). *Executing Data Quality Projects.* Morgan Kaufmann.

McKinsey & Company. (2017, September 07). *Human + machine: A new era of automation in manufacturing.* Retrieved from McKinsey & Company: https://www.mckinsey.com/capabilities/operations/our-insights/human-plus-machine-a-new-era-of-automation-in-manufacturing

McKinsey Global Institute. (2015, December 01). *Digital America: A tale of the haves and have-mores.* Retrieved from McKinsey & Company: https://www.mckinsey.com/industries/technology-media-and-telecommunications/our-insights/digital-america-a-tale-of-the-haves-and-have-mores

McKinsey Global Institute. (2017). *Jobs lost, jobs gained: What the future of work will mean for jobs, skills, and wages.* McKinsey & Company.

McSweeney, K. (2020, January 31). *NOW Powered by Northrop Grumman.* Retrieved from Northrop Grumman:

https://now.northropgrumman.com/ray-kurzweil-predictions-persist-turns-70/

Megginson, L. (1963). Lessons from Europe for American Business. *Southwestern Social Science Quarterly*, 44(1): 3-13, at p. 4.

Metz, C., & Weise, K. (2023, January 12). *Microsoft Bets Big on the Creator of ChatGPT in Race to Dominate A.I.* Retrieved from The New York Times: https://www.nytimes.com/2023/01/12/technology/microsoft-openai-chatgpt.html

Metz, C., & Weise, K. (2023, January 23). *Microsoft to Invest $10 Billion in OpenAI, the Creator of ChatGPT.* Retrieved from The New York Times: https://www.nytimes.com/2023/01/23/business/microsoft-chatgpt-artificial-intelligence.html

Meyersohn, N. (2018, March 06). *Why Domino's is winning the pizza wars.* Retrieved from CNN Money: https://money.cnn.com/2018/03/06/news/companies/dominos-pizza-hut-papa-johns/index.html

Naftulin, J. (2016, Juy 13). *Here's how many times we touch our phones every day.* Retrieved from Business Insider: https://www.businessinsider.com/dscout-research-people-touch-cell-phones-2617-times-a-day-2016-7

NAIC. (2022, March 4). *NAIC - Accelerated Underwriting (A) Working Group Draft.* Retrieved from NAIC Life insurance and Annuities Committee:

https://content.naic.org/sites/default/files/inline-files/AUWG%20DRAFT%203-4-22%20for%20SNM_0.pdf

NASA. (n.d.). *Europa Clipper Mission.* Retrieved from NASA: https://www.jpl.nasa.gov/missions/europa-clipper

Neill, B., Hallmark, J. D., Jackson, R. J., & Diasio, D. (2023, October 31). *Key takeaways from the Biden administration executive order on AI.* Retrieved from EY: https://www.ey.com/en_us/public-policy/key-takeaways-from-the-biden-administration-executive-order-on-ai

Neuburger, J. D. (2023, February 23). *ChatGPT Risks and the Need for Corporate Policies.* Retrieved from National Law Review: https://natlawreview.com/chatpgpt/risks-and-need-corporate-policies

Neufield, D. (2021, November 08). *There are 1.8 billion millennials on earth. Here's where they live.* Retrieved from World Economic Forum : https://www.weforum.org/agenda/2021/11/millennials-world-regional-breakdown/

Neuman, S. (2021, September 14). *Scientists Say They Could Bring Back Woolly Mammoths. But Maybe They Shouldn't.* Retrieved from National Public Radio (NPR.org): https://www.npr.org/2021/09/14/1036884561/dna-resurrection-jurassic-park-woolly-mammoth

NIH. (2019, 03 12). *Computers, Artificial Intelligence, and Expert Systems in Biomedical Research.* Retrieved from National

Institute of Health (NIH) - National Library of Medicine: Profiles in Science - The Joshua Lederberg Papers: https://profiles.nlm.nih.gov/spotlight/bb/feature/ai

Niiler, E. (2019, January 25). *How the Second Industrial Revolution Changed Americans' Lives*. Retrieved from The History Channel: https://www.history.com/news/second-industrial-revolution-advances

Nikkei. (2021, July 09). *JAL aims to launch flying car business in fiscal 2025*. Retrieved from Nikkei Asia: https://asia.nikkei.com/Business/Technology/JAL-aims-to-launch-flying-car-business-in-fiscal-2025

NIST. (2022, May 05). *Software Supply Chain Security Guidance*. Retrieved from National Institute of Standards and Technology: https://www.nist.gov/itl/executive-order-14028-improving-nations-cybersecurity/software-supply-chain-security-guidance

Nokiamob. (2017, December 11). *Can Anyone Catch Nokia*. Retrieved from Nokiamob: https://nokiamob.net/2017/11/12/forbes-in-2007-can-anyone-catch-nokia/

NPR. (2021). *NPR Special Series - The CRISPR Revolution*. Retrieved from National Public Radio (NPR): https://www.npr.org/series/773368439/the-crispr-revolution

NYT. (1973, 09 13). *Mrs. Marjorie Merriweather Post Is Dead at 86*. Retrieved from The New York Times:

https://www.nytimes.com/1973/09/13/archives/mrs-marjorie-merriweather-post-is-dead-at-86-a-rich-working-woman.html

NYT. (2022, January 3). *NYT Technology*. Retrieved from The New York Times: https://www.nytimes.com/2022/01/03/technology/personaltech/blackberry-devices-stop-working.html

OpenAI. (2021). *OpenAI Codex*. Retrieved from OpenAI Codex: https://openai.com/

OSHA. (2023). *Agricultural Operations*. Retrieved from United States Department of Labor - Occupational Safety and Health Administration: https://www.osha.gov/agricultural-operations

Overly, S. (2016, July 21). *Humans once opposed coffee and refrigeration. Here's why we often hate new stuff.* Retrieved from The Washington Post: https://www.washingtonpost.com/news/innovations/wp/2016/07/21/humans-once-opposed-coffee-and-refrigeration-heres-why-we-often-hate-new-stuff/

Owens, E., Sheehan, B., Mullins, M., Cunneen, M., Ressel, J., & Castignani, G. (2022). *Explainable Artificial Intelligence (XAI) in Insurance.* MDPI.

(2011). *Oxford Dictionaries.* Oxford University Press.

Pande, P. (2021, July 12). *JAL Looks At eVTOL Operations From 2025.* Retrieved from Simple Flying: https://simpleflying.com/jal-evtol-2025/

Panetta, K. (2021, August 26). *A Data and Analytics Leader's Guide to Data Literacy.* Retrieved from Gartner: https://www.gartner.com/smarterwithgartner/a-data-and-analytics-leaders-guide-to-data-literacy#:~:text=Gartner%20defines%20data%20literacy%20as,case%2C%20application%20and%20resulting%20value.

Pantic, N. (2021). *How Many Photos Will Be Taken in 2021?* Retrieved from mylio: https://blog.mylio.com/how-many-photos-will-be-taken-in-2021-stats/

Perifanis, N.-A. a. (2023, 02). Investigating the Influence of Artificial Intelligence on Business Value in the Digital Era of Strategy: A Literature Review. *Information Vol. 14*, p. 85.

Pew. (2015). *Labor Force Composition by Generation.* PEW RESEARCH CENTER.

Phillips, P. J., Hahn, C. A., Fontana, P. C., Yates, A. N., & Greene, K. (2021). *Four Principles of Explainable Artificial Intelligence.* Gaithersburg, MD: NIST.

Phillpott, S. (2021, May 17). *15 Disappearing Jobs that Won't Exist in 2030.* Retrieved from CareerAddict: https://www.careeraddict.com/disappearing-jobs

Powell, A. (2020, November 11). *AI revolution in medicine.* Retrieved from The Harvard Gazette: https://news.harvard.edu/gazette/story/2020/11/risks-and-benefits-of-an-ai-revolution-in-medicine/

Profit&. (2019, September 20). *7 REAL-LIFE EXAMPLES OF SUCCESSFUL CHANGE MANAGEMENT IN BUSINESS.* Retrieved from Profit&: https://insights.profitand.com/blog/real-life-examples-of-successful-change-management-in-business

Purdue. (2020). *Generational Differences in the Workplace.* Retrieved from Purdue University Global: https://www.purdueglobal.edu/education-partnerships/generational-workforce-differences-infographic/

Qlik. (n.d.). *Qlik.* Retrieved from Qlik: https://www.qlik.com/us/data-literacy#:~:text=It's%20a%20skill%20that%20empowers,and%20communicate%20meaning%20to%20others.

Rao, A. (2023). *Sizing the prize - PwC's Global Artificial Intelligence Study: Exploiting the AI Revolution.* Retrieved from PwC: https://www.pwc.com/gx/en/issues/data-and-analytics/publications/artificial-intelligence-study.html

Reddy, C. (2017, October 05). *Kurzweil Claims That the Singularity Will Happen by 2045.* Retrieved from Futurism.com: https://futurism.com/kurzweil-claims-that-the-singularity-will-happen-by-2045

Reddy, C. (2017, October 05). *Kurzweil Claims That the Singularity Will Happen by 2045.* Retrieved from Futurism.com: https://futurism.com/kurzweil-claims-that-the-singularity-will-happen-by-2045

Redman, T. (n.d.). *Why is Data Quality Important with Tom Redman.* Retrieved from AtScale : https://www.atscale.com/podcast/thomas-redman-s1e7/

Regalbuto, J. (2019). *Use of External Consumer Data and Information Sources in Underwriting for Life Insurance.* Albany, NY: New York Department of Financial Services.

Richards, A. (2012, May 18). *Clarence Birdseye And His Fantastic Frozen Food Machine.* Retrieved from National Public Radio: https://www.npr.org/sections/thesalt/2012/05/18/152743718/clarence-birdseye-and-his-fantastic-frozen-food-machine

Roose, K. (2023, February 3). *How ChatGPT Kicked Off an A.I. Arms Race.* Retrieved from The New York Times: https://www.nytimes.com/2023/02/03/technology/chatgpt-openai-artificial-intelligence.html

Roser, M. (2023, February 07). *AI timelines: What do experts in artificial intelligence expect for the future?* Retrieved from Our World in Data: https://ourworldindata.org/ai-timelines

Roser, M., Ortiz-Ospina, E., & Ritchie, H. (2019). *Life Expectancy (United Nations, Our World In Data).* United Nations.

Rosso, C. (2023, December 24). *AI Discovers First New Antibiotic in Over 60 Years.* Retrieved from Psychology Today: https://www.psychologytoday.com/us/blog/the-future-brain/202312/ai-discovers-first-new-antibiotic-in-over-60-years

Sakthivel, K. (2021). *Themes, Trends, and Considerations Shaping the Future of Technology in Financial Services - The Emerging Technology Series.* Windsor CT: LIMRA.

Sathe, M., & Ruloff, K. (2023, July 28). *The EU AI Act: What it means for your business.* Retrieved from EY Switzerland: https://www.ey.com/en_ch/forensic-integrity-services/the-eu-ai-act-what-it-means-for-your-business

Sayre, M. A. (2022). Life's Not Fair. Is Life Insurance? *Privacy Certificate Student Publications.*

Schwab, K. (2015, December 12). *The Fourth Industrial Revolution - What It Means and How to Respond.* Retrieved from Foreign Affairs: https://www.foreignaffairs.com/articles/2015-12-12/fourth-industrial-revolution

Schwab, K. (2017). *The Fourth Industrial Revolutuion.* Currency.

Scientist, N. (2021). *New Scientist.* Retrieved from New Scientist: https://www.newscientist.com/definition/what-is-crispr/

Sebastian-Coleman, L. (2022). *Meeting the Challenges of Data Quality Management* . Academic Press.

Siau, K., & Wang, W. (2018). Building trust in artificial intelligence, machine learning, and robotics. *Cutter Business Technology* , 47-53.

Snider, M., & Molina, B. (2021, November 10). *Everyone wants to own the metaverse including Facebook and Microsoft. But what exactly is it?* Retrieved from USA Today:

https://www.usatoday.com/story/tech/2021/11/10/metaverse-what-is-it-explained-facebook-microsoft-meta-vr/6337635001/

Soberanis, L. (2021, September 13). *From underwriting to claims management, artificial intelligence will transform the insurance industry.* Retrieved from IBM: https://www.ibm.com/blogs/watson/2021/09/ai-will-transform-insurance-industry/

Soni, K. (2022, April 04). *Data governance framework set up by Uber.* Retrieved from CIO News: https://cionews.co.in/data-governance-framework-set-up-by-uber/

Stein, R. (2021, June 26). *National Public Radio - All Things Considered.* Retrieved from National Public Radio (NPR): https://www.npr.org/sections/health-shots/2021/06/26/1009817539/he-inherited-a-devastating-disease-a-crispr-gene-editing-breakthrough-stopped-it

study.com. (2014, February 06). *The Second Industrial Revolution: Timeline & Inventions.* Retrieved from Study.com: https://study.com/academy/lesson/the-second-industrial-revolution-timeline-inventions.html

Surowiecki, J. (2013, September 03). *Where Nokia Went Wrong.* Retrieved from The New Yorker: https://www.newyorker.com/business/currency/where-nokia-went-wrong

Susarla, A. (2023, November 04). *Analysis: How Biden's new executive order tackles AI risks, and where it falls short.* Retrieved from PBS: https://www.pbs.org/newshour/politics/analysis-how-bidens-new-executive-order-tackles-ai-risks-and-where-it-falls-short

Sweden, E. (2009). *Data Governance Part II: Maturity Models – A Path to Progress.* Lexington, KY: NASCIO.

Takyar, A. (2022). *AI USE CASES & APPLICATIONS ACROSS MAJOR INDUSTRIES* . Retrieved from LeewayHertz: https://www.leewayhertz.com/ai-use-cases-and-applications/#:~:text=9.-,Manufacturing,and%20adoption%20of%20AI%20technology.

Terrell, K. (2022, June 02). *10 Jobs That Are Disappearing Fast.* Retrieved from AARP: https://www.aarp.org/work/careers/disappearing-jobs/

Tewksbury, K. (2022, August). Chasing Amazon: Accelerated Underwriting Focuses on CX. *LIMRA MarketFacts.*

The Biden Administration. (2023, October 30). *FACT SHEET: President Biden Issues Executive Order on Safe, Secure, and Trustworthy Artificial Intelligence.* Retrieved from The White House: https://www.whitehouse.gov/briefing-room/statements-releases/2023/10/30/fact-sheet-president-biden-issues-executive-order-on-safe-secure-and-trustworthy-artificial-intelligence/

TKSST. (2012). *Carl Sagan's Cosmic Calendar.* Retrieved from The Kid Should See This - Smart videos for curious minds of all ages: https://thekidshouldseethis.com/post/cosmic-calendar-carl-sagan-video

Tocci, M. (2020). *What Pleasures Would You Sacrifice to Keep Your Smartphone?* SimpleTexting.

Toreini, E., Aitken, M., Coopamootoo, K., Elliott, K., Carlos Gonzalez Zelaya, & Moorsel, A. V. (2020). The relationship between trust in AI and trustworthy machine learning technologies. *Conference on Fairness, Accountability, and Transparency.* Barcelona, Spain.

Turek, M. (2017). *Explainable Artificial Intelligence (XAI).* Retrieved from DARPA: https://www.darpa.mil/program/explainable-artificial-intelligence

UALR. (2018, April 26). *UA Little Rock Data Governance Policy Framework.* Retrieved from University of Arkansas at Little Rock: https://ualr.edu/institutionalresearch/files/2018/11/UA-Little-Rock-Data-Governance-Policy-Framework-for-UA-Little-Rock-2-22-18.pdf

USDA. (2023). *About USDA.* Retrieved from United States Department of Agriculture: https://www.usda.gov/our-agency/about-usda/laws-and-regulations

Vega, N. (2021, December 27). *Microsoft's market cap grew more than $800 billion in 2021—here's how it compares to the* *most valuable companies in the world.* Retrieved from CNBC: https://www.cnbc.com/2021/12/27/how-much-the-biggest-companies-grew-in-2021.html

Verisk. (2020, August 10). *Verisk Launches EHR Triage Engine to Help Speed Approval of Life Insurance Applications.* Retrieved from Verisk: https://www.verisk.com/newsroom/verisk-launches-ehr-triage-engine-to-help-speed-approval-of-life-insurance-applications/

Wall, M. (2021, July 7). *Enceladus.* Retrieved from Space.com: https://www.space.com/methane-plume-enceladus-possible-sign-alien-life

Wang, W., Wang, J., Chen, C., Su, S., Chu, C., & Chen, G. (2022). *A Capability Maturity Model for Intelligent Manufacturing in Chair Industry Enterprises.* Processes.

Watson, H. (2020). The Need for Explainable Processes and Algorithms. *Business Intelligence Journal.*

Weidner, J. B. (2021, August 8). *How and Why Google Glass Failed.* Retrieved from Investopedia: https://www.investopedia.com/articles/investing/052115/how-why-google-glass-failed.asp

Weitz, A. (2017). *The Camera Phone is 20 Years Old!* Retrieved from B&H: https://www.bhphotovideo.com/explora/photography/features/camera-phone-20-years-old

Wikipedia. (2022). *iPhone.* Retrieved from Wikipedia: https://en.wikipedia.org/wiki/IPhone

Wikipedia. (n.d.). *USB.* Retrieved from Wikipedia: https://en.wikipedia.org/wiki/USB

Wikipedia. (n.d.). *Wikipedia.* Retrieved from Clarence Birdseye: https://en.wikipedia.org/wiki/Clarence_Birdseye

Wilkins, A. (2012, March 25). *The last mammoths died out just 3600 years ago...but they should have survived.* Retrieved from Gizmodo: https://gizmodo.com/the-last-mammoths-died-out-just-3600-years-ago-but-th-5896262

Wilkinson, D. (2020). *2020 SMS Marketing Report.* SimpleTexting.

Witten, S. (2019, November 08). *The death of the DVD: Why sales dropped more than 86% in 13 years.* Retrieved from CNBC: https://www.cnbc.com/2019/11/08/the-death-of-the-dvd-why-sales-dropped-more-than-86percent-in-13-years.html

Zurich. (2022, June 24). *How will Gen Z change the workplace?* Retrieved from Zurich: https://www.zurich.com/en/media/magazine/2022/how-will-gen-z-change-the-future-of-work#:~:text=Gen%20Z%20could%20change%20jobs,a%20survey%20of%20U.S.%20students.

About the Author

Kartik Sakthivel $_{PhD}$ serves as the **Vice President and global Chief Information Officer** for LIMRA LOMA and LL Global, a non-profit worldwide trade association for the financial services industry. Kartik has had a vast amount of leadership and management experience across multiple industries. He has held leadership roles in companies of all sizes, ranging from small and midsize firms to Fortune 100 organizations.

A technology practitioner, Kartik has been urging industries and workers within the industries to adopt and adapt to the massive digital disruption that is the hallmark of the early 21st century. He delivered a **TED Talk** at TEDx Portsmouth in Portsmouth, New Hampshire in September of 2019, advocating for individuals to embrace technology to thrive in their professions. Additionally, believing that encouraging industries and people to continually invent and reinvent themselves is an act of consummate leadership, Kartik is on a mission to impress upon leaders how crucial they are to the success of their organizations and their people. With an immutable belief that leadership is not about a title, but how you serve your teams, Kartik has been coaching and mentoring new leaders on how they can start the process of leaving their leadership legacy. Kartik has published two other books in addition to DNAI. "**Find Your Red Cape**" is a book about leadership, and provides a framework for leaders to discover their "leadership superpowers". "**Digital Planet, Human Inhabitants**" is an extended version of his TED Talk, and provides individuals with the guidance and advice they need to demonstrate success in the Digital Age.

Kartik Sakthivel

Kartik holds a **PhD** in **Computer and Information Science** (with a focus on Information Quality and a dissertation on best practices for **AI**) and a **Graduate Certificate in Information Quality** from University of Arkansas at Little Rock, a **Master's Degree in Computer Information Systems/IT**, and a **Master of Business Administration** degree from Southern New Hampshire University.

Born in Mumbai (Bombay) India, Kartik has been a resident of Dover, New Hampshire in the United States of America for his entire adult life.

DNAI

Made in the USA
Monee, IL
05 November 2026

67815690R00315